# Guitar Amplifier Electronics
# Fender Deluxe®

Richard Kuehnel

Amp Books®
Seattle
www.ampbooks.com

Printed in the United States of America

This work is sold with the understanding that the author and publisher are not engaged in rendering any professional services. If professional assistance is required, the services of a competent professional person should be sought. Nothing in this book, including the accuracy of the information it contains, is guaranteed for any particular purpose. The book is presented "as is" and the author and publisher do not offer any expressed or implied warranties or representations, nor accept any liability and responsibility to any person or entity for damage or loss caused, or alleged to be caused, directly or indirectly, by the use, misuse, negligent use, or inability to use information contained in this book.

BANDMASTER®, BASSMAN®, CHAMP®, CONCERT®, DELUXE®, 57 DELUXE®, HARVARD®, PRINCETON®, PRO®, SHOWMAN®, SUPER®, TREMOLUX®, TWIN®, TWIN REVERB®, VIBRASONIC®, VIBROLUX®, and VIBROVERB® are trademarks of Fender Musical Instruments Corporation (FMIC). All rights reserved. FMIC is not affiliated in any way with the author or publisher.

ISBN 978-0-9996742-4-6

2nd Printing

Copyright © 2021 by Richard Kuehnel. All rights reserved.

No part of this publication may be reproduced, stored in a retrieval system, or transmitted, in any form or by any means, electronic, mechanical, photocopying, recording, or otherwise, without the prior written permission of the author.

# Contents

Introduction ........................................................................................... 1
   Warning: High Voltage! ..................................................................... 1
   Deluxe versus Deluxe ........................................................................ 1
   Evolutionary Development ................................................................ 2
   Acknowledgements ............................................................................ 4
Chapter 1: Heritage – The Woody and Early Tweeds ........................... 5
   Model 26 Woody ................................................................................ 5
      System Design Concept ................................................................. 7
      6V6 Push-Pull Power Amp ........................................................... 8
      5Y3 Power Supply ......................................................................... 9
      6N7 Paraphase Phase Inverter .................................................... 10
      Volume Controls ........................................................................... 13
      6SC7 Preamp ................................................................................ 17
      Audio Signal Levels ...................................................................... 19
   Model 5A3 TV Front ......................................................................... 20
      System Design Concept ............................................................... 22
      Power Amp and Power Supply Adjustments ............................ 23
      A New Phase Inverter Tube ......................................................... 24
      Integrating Volume and Tone Controls .................................... 25
      Introducing Grid-Leak Bias ......................................................... 27
      Audio Signal Levels ...................................................................... 29
   Model 5B3 Wide Panel ..................................................................... 31
   Model 5C3 Wide Panel ..................................................................... 35
      System Design Concept ............................................................... 35
      Preamp Tweaks ............................................................................ 37
      Negative Feedback ....................................................................... 38
Chapter 2: Legacy – The Wide Panel Deluxe 5D3 ............................... 39
   System Design Concept .................................................................... 39
   Power Amp Performance ................................................................. 42
   Power Supply Ripple ........................................................................ 43
   Introducing a Self-Balancing Paraphase Phase Inverter ................ 44

| | |
|---|---|
| Volume and Tone Control Insertion Loss | 48 |
| A New Preamp Design | 48 |
| System Profile | 53 |
| System Voicing | 55 |
| Evolutionary Development Continues | 55 |
| **Chapter 3: Pinnacle – The Narrow Panel Deluxe 5E3** | **57** |
| System Design Concept | 57 |
| Coaxing More Power from the Power Amp | 60 |
| Power Amp Performance | 62 |
| Power Supply Ripple | 64 |
| A New Voltage Amplifier and Split-Load Phase Inverter | 65 |
|     Voltage Amplifier Performance | 66 |
|     Phase Inverter Performance | 67 |
| Volume and Tone Control Interactivity | 69 |
| Preamp Adjustments | 72 |
| Substituting a 12AX7 for the 12AY7 | 73 |
| A New, Iconic Front End | 76 |
| System Profile | 78 |
| System Voicing | 80 |
| Full Power Performance – Class A Operation? | 81 |
| Harmonic Distortion at Full Power | 84 |
| Intermodulation Distortion (IMD) | 88 |
| Overdrive Dynamics | 90 |
|     No Feedback, Maximum Grunge | 91 |
|     Phase Inverter Effects | 91 |
|     Bias Excursion and Power Supply Sag | 96 |
| A Pinnacle and a Pivot | 98 |
| **Chapter 4: Enhancement – The Brownface Deluxe 6G3** | **99** |
| System Design Concept | 99 |
| Coaxing Even More Power from the Power Amp | 102 |
| Power Amp Performance | 102 |
| Power Supply Modifications | 105 |

Introducing a DC Bias Supply .......................................................... 108

Introducing a Long-Tailed-Pair Phase Inverter ............................... 109

Introducing Negative Feedback ....................................................... 114

How Negative Feedback Reduces Distortion ................................... 116

Feedback Calculations for the 6G3 .................................................. 117

An Additional Voltage Amplification Stage ..................................... 120

More Aggressive Volume and Tone Controls ................................... 122

Increasing First-Stage Gain ............................................................. 124

Introducing Tremolo ........................................................................ 129

System Profile .................................................................................. 133

System Voicing ................................................................................ 135

Full Power Performance .................................................................. 136

Bias Excursion and Power Supply Sag ............................................. 138

Chapter 5: Perfection – The Blackface Deluxe Reverb ....................... 141

AA763 System Design Concept ....................................................... 141

Coaxing Even More Power from the Power Amp ........................... 145

Screen Voltage Swing at Full Power ................................................ 146

Power Supply Modifications ........................................................... 149

Introducing Signal-Modulation Tremolo ......................................... 150

Modifying the LTP Phase Inverter ................................................... 154

Increasing Negative Feedback ......................................................... 157

Second-Stage Voltage Amplifiers .................................................... 160

First Stage, Tone Stack, and Volume Control .................................. 162

System Profile .................................................................................. 165

System Voicing ................................................................................ 168

Full Power Performance .................................................................. 169

Bias Excursion and Power Supply Sag ............................................. 171

Modifications Incorporated into the AB763 .................................... 172

    Tweaking the Tremolo, Power Amp, and Phase Inverter ............... 173

    Tweaking the Tone Stack ............................................................. 175

Reverb Design Concept ................................................................... 176

Reverb Recovery and Mixer Amplifiers .......................................... 180

Reverb Tank ................................................................................................ 184
Reverb Output Transformer ......................................................... 186
Parallel 12AT7 Reverb Driver........................................................ 187
Some Final Words on the AB763 Fender Deluxe Reverb .................. 190
Appendix ....................................................................................... 192
Decibel-Volts (dBV) ..................................................................... 192
AB763 Supply Voltage Calculations.............................................. 192
Index .............................................................................................. 194

# Introduction

## Warning: High Voltage!

Marom Bikson of the Department of Biomedical Engineering, City University of New York, reports that as little as 25 volts can be lethal under some circumstances.[1] Guitar amplifier circuits use hundreds of volts, so please heed the following warning.

> Vacuum tube circuits generally use lethally high voltages. Capacitors in the circuit store these deadly voltages long after the amplifier has been turned off and unplugged. Please do not work on one of these circuits unless you are properly trained.

## Deluxe versus Deluxe

In 1954 Elvis Presley recorded "That's alright, Mama" at Sam Phillips' studios in Memphis, Tennessee. More than two decades later, Shelter Records produced Tom Petty and the Heartbreakers' "Breakdown." Whether it's the mids-heavy fang of Scotty Moore's double stops or the focused magnetism of Mike Campbell's blues riffs, there is no mistaking the unmitigated glory of a well-tempered guitar plugged into a late 1950's tweed Deluxe. Guitar World calls it one of the most iconic guitar amps of all time.

> "Introduced in 1948, the Fender Deluxe was praised for its dynamic, harmonically rich overdrive and compression. It was offered in numerous configurations and designs over the years, but the most desirable model is the 5E3 narrow-panel Deluxe, built from 1955 to 1960 and offered in a tweed-covered cabinet."[2]

---

[1] Marom Bikson, "A Review of Hazards Associated with Exposure to Low Voltages." Available at:
http://bme.ccny.cuny.edu/faculty/mbikson/BiksonMSafeVoltageReview.pdf (Accessed on January 28, 2018.)

[2] Damian Fanelli, Christopher Scapelliti and Tom Gilbert, "The 10 Most Iconic Guitar Amps of All Time," **Guitar World**, April 24, 2020.

There is no bigger fan of the 5E3 than Neil Young.

> "At the core of Young's amplifier setup is a piece of gear as essential to his sound as [his 1953 Gibson Les Paul goldtop] Old Black: the 1959 tweed Fender Deluxe he's used since the late Sixties. A small, 15-watt unit, with just two volume knobs and a shared tone control, this amp, says [Young's guitar tech Larry] Cragg, "makes all the sound. Onstage, as loud as everything gets, that's what you hear."[3]

The 5E3 is the most cloned amp of all time. Its elegantly straightforward signal chain maximizes tone and minimizes parts count, making it a favorite for kit makers and kit builders alike. In fact, there is perhaps only one other amp design on the entire planet that can compete with the 5E3 in terms of popularity, particularly among working musicians. Ironically, it also carries the name "Deluxe," the AB763 Deluxe Reverb.

> "A Fender Deluxe Reverb is usually what I use, because it's high enough wattage to where the low tunings don't break up too much."[4] –Madison Cunningham

> "In 1993 or '94 I bought a '64 Deluxe Reverb and to this day it's my favorite. We have tons of amps sent to us, and we both love that one."[5] –Susan Tedeschi

> "Roy Buchanan and his trusty, well-weathered 50's telecaster never abused a finer vintage amp than the Fender Deluxe Reverb.[6]

> "If I had to have one amp to be stuck with on the desert island, give me a Deluxe Reverb.[7] –Aspen Pittman

Whether you prefer a narrow panel 5E3 or a blackface AB763 Reverb, the sonic palette of a vintage Fender Deluxe is an unmistakable pleasure to engage. This book examines both electronic circuits in great detail.

## Evolutionary Development

New product development is a process that is accelerated by a legacy of successful products. Here is a modern example.

> "Representing the fourth collaboration between Dr. Z and Brad Paisley, the DB4 is the outcome of the country star's request for a rich and warm

---

[3] Richard Bienstock, ""Neil Young: Ragged Glory," **Guitar World**, October 15, 2009.
[4] Jim Beaugez, "Beyond Standard," **Guitar Player**, June 2020, p. 21.
[5] Ward Meeker, "Susan Tedeschi and Derek Trucks, Minds Made Up," **Vintage Guitar**, October 2013.
[6] Ben Fargen, "10 Classic Guitar Amps & the Songs that Made Them Famous," **My Rare Guitars**, January 1, 2012.
[7] Tom Wheeler, **The Soul of Tone**, (Milwaukee: Hal Leonard, 2007), pp. 479-480.

sounding British-voiced amp. Dr. Z founder Mike Zaite says he started from scratch with this design, deciding early on a U.S.-made N.O.S. 5879 pentode tube for the preamp."[8]

The subtext here is that most professional designs are not started "from scratch." Instead, they *evolve* from previous designs.[9] Likewise, the Deluxe 5E3 represents evolutionary development based on a history of modifications to a basic architecture. The 5A3/5B3 takes the Deluxe Model 26 and gives it grid-leak bias, a new phase inverter tube, and higher DC supply voltages. The 5C3 applies a smidgeon of negative feedback.[10] The 5D3 reverts back to cathode bias, eliminates the feedback, and transitions to 9-pin triodes. The 5E3 tweaks the design further.

Deluxe Evolution 1946-1965

Each of these system modifications is made in the context of similar changes to the rest of Fender's product line. From the 5C3 to the 5E3, for example, the Deluxe gets a self-balancing paraphase and then a split-load phase inverter.[11] The Fender Super takes the same design path from the 5C4 to the 5E4.

---

[8] Art Thompson, "Dr. Z DB4," **Guitar Player**, August 2016, p. 74.
[9] Richard Kuehnel, **Fundamentals of Guitar Amplifier System Design**, (Seattle: Amp Books, 2019), p. 16.
[10] Richard Kuehnel, **Guitar Amplifier Electronics: Basic Theory**, (Seattle: Amp Books, 2018), pp. 142-153.
[11] **Basic Theory**, pp. 130-134.

Fender's product development represents a master class in evolutionary design. For this reason, we begin our circuit exploration with the Deluxe Model 26, one of Fender's original "woodies," and chart the complete heritage of the 5E3 circuit. We pay particular attention to the wide-panel 5D3, which serves as the baseline from which the narrow-panel 5E3 is crafted. We also examine the brownface 6G3 in depth, because it is a major milestone in the evolutionary development of the blackface AB763 Reverb.

Vintage Deluxe circuit development is a story of heritage, legacy, refinement, and enhancement in a quest for musical perfection. Welcome to the journey.

## Acknowledgements

I want to express my appreciation to Paul Reid for technical review. I also want to thank my wife Tipi, to whom this book is dedicated, for her unfailing encouragement throughout the preparation of this work.

# Chapter 1: Heritage – The Woody and Early Tweeds

The narrow panel tweed 5E3 and the blackface AB763 Reverb are the most famous of the Fender's Deluxe models. For this reason, we dedicate entire chapters to these systems. In this chapter we examine their heritage in the context of Fender's newly evolving product line.

## Model 26 Woody

Prior to the Deluxe, Leo Fender worked with Doc Kauffman on a series of K&F amplifiers. They were very simple designs by today's standards, using tubes like the 6SF5, a metal tube with only a single triode. Some of these early amps came with no controls.

> "Why build an amp with no tone or volume controls? That's what the knobs on your guitar were for. Remember, Leo Fender saw the amp as an integral part of the guitar. In his mind, the "instrument" was the guitar, cord, and amp."[12] –Tom Wheeler

The first Deluxe to be called a "Fender" was the Model 26, one of three "woodies" Leo created in 1946.[13]

> "The first Fender brand amps are nicknamed 'woodies' because of their hardwood cabinets and matching handles. According to Richard Smith, Leo Fender received a shipment of hardwood intended for steel guitar bodies in 1946, but the one-inch-thick pieces were too thin for their intended use." "As it sat in an empty lot, Leo designed the first line of Fender amps in order to use it before termites could eat it."[14]

From the very beginning the Deluxe occupied the middle of Fender's product line – the Model 26 with one 10-inch speaker stood halfway between the 6-watt, 1x8 Princeton student amp and the 6-tube, 1x15 Professional.[15] Fender advertising from 1947 claims an output of 14 watts, which matches a similar power amp design from a 6V6 data sheet.

It might be said that subsequent Deluxe models evolve from this "original," but even the first Fender Deluxe has a heritage – in the mid-1940s its triode voltage amplifier, paraphase phase inverter,[16] and push-pull power amp were familiar audio circuits.

---

[12] Tom Wheeler, **The Soul of Tone**, (Milwaukee: Hal Leonard, 2007), p. 113.
[13] John Teagle and John Sprung, **Fender Amps: The First Fifty Years**, (Milwaukee: Hal Leonard, 1995), p. 25.
[14] Tom Wheeler, p. 119.
[15] Tom Wheeler, p. 120.
[16] Richard Kuehnel, **Guitar Amplifier Electronics: Basic Theory**, (Seattle: Amp Books, 2018), pp. 126-130.

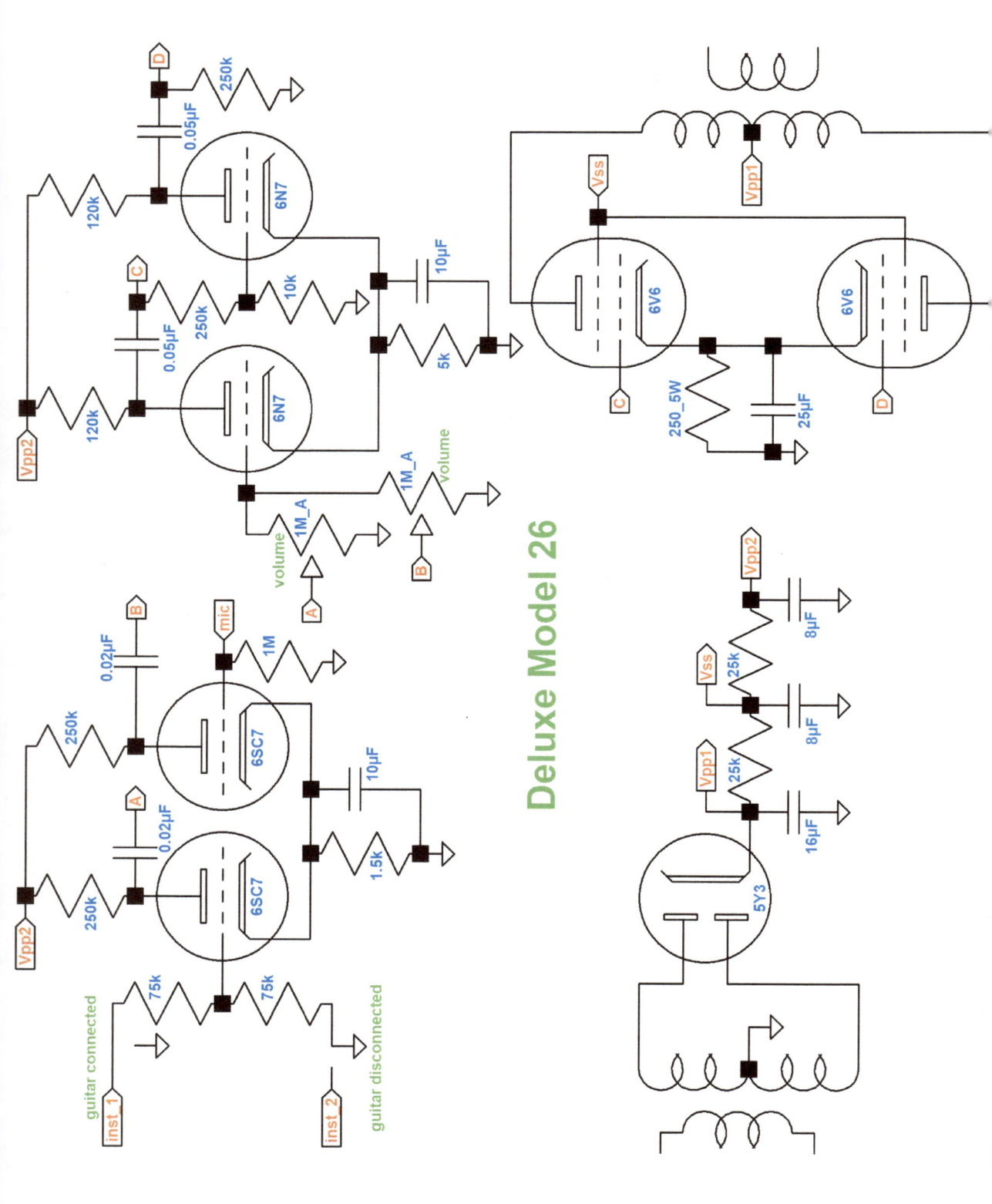

"His circuits were 'cookbook' designs, AT&T/Western Electric circuits printed in tube manuals. He borrowed heavily. Nothing wrong with that. Hell, some of my designs were cookbook, too. Mr. Fender's brilliance was in the way he established the template for guitar amplifiers, and for 19 years he out-innovated the rest of this entire industry."[17] – Hartley Peavey

With a two-guitar instrument channel and a microphone channel, each with its own volume control, the first Deluxe was quite versatile for a 1940s guitar amp.[18] By today's standards, on the other hand, the Model 26 is a bare-bones design.

## System Design Concept

The signal path consists of a 6SC7 voltage amplifier, a volume control circuit, a 6N7 phase inverter, and a 6V6 push-pull power amp. For the DC plate and screen supplies, a 5Y3 vacuum tube rectifier drives a capacitor input and two RC ripple filters.[19]

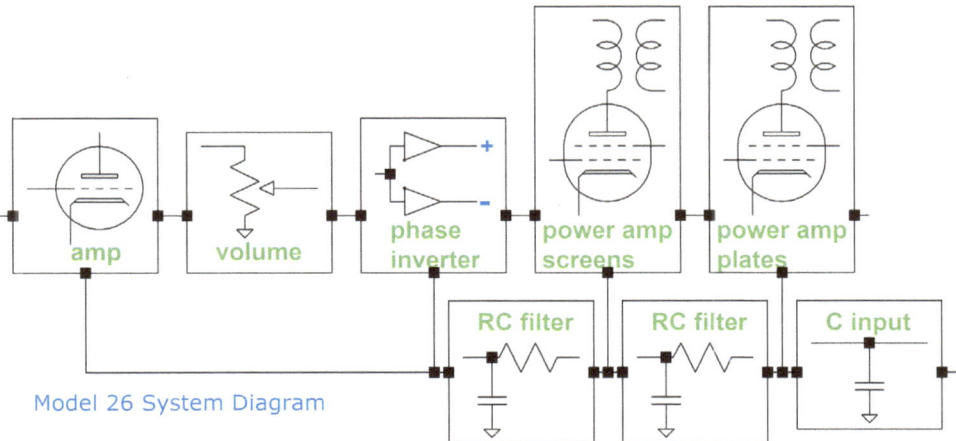

Model 26 System Diagram

As a system, the design is very simple. In the 1940s a guitar amplifier was first and foremost an *amplifier* designed to boost the weak signal from a guitar pickup to a power level capable of driving a *loudspeaker*.

"One of the beautiful things about these Fender circuits is their simplicity. In my experience as a designer, we prototype all sorts of elaborate ideas,

---

[17] Tom Wheeler, **The Soul of Tone**, (Milwaukee: Hal Leonard, 2007), p. 27.
[18] John Teagle and John Sprung, **Fender Amps: The First Fifty Years**, (Milwaukee: Hal Leonard, 1995), p. 42.
[19] Richard Kuehnel, **Guitar Amplifier Electronics: Basic Theory**, (Seattle: Amp Books, 2018), pp. 160-161.

and they sound okay, but you find as you take things away the sound almost always gets better and better."[20] –Steve Carr

With octal tubes outside and point-to-point wiring inside, the chassis has a solid, public-address-system vibe, which may stem from the influence of Ray Massie, a repairman at Fender Radio Service. Don Randall, Leo Fender's distributor at Radio & Television Equipment Company, describes the situation this way.[21]

> "I think Ray had a lot to do with the early circuits because he was in that business to begin with. Leo was coming to it out of his repair business and his work with public address systems and so on. Ray knew a lot about circuitry, and Leo didn't know that much about it at the very beginning. Of course, he learned everything eventually. But Ray really knew that stuff. I think he knew more about circuits than Leo did when they started. I wouldn't say that he was actually designing products, but I think he had a big hand in helping with some of the technical details and so forth."[22]

The signal paths for the Model 26 microphone and guitar inputs are identical. System voicing is the epitome of minimalism – with the volume control at noon the amp has a completely flat response from the input jack to the speaker. A treble cut control across the output transformer primary (not shown and not recommended for safety reasons) is the only nod to frequency shaping.

As a system, the Model 26 is elegant in its lack of complexity. It is the first dollop of paint on the Deluxe canvas, the humble beginnings of sonic masterpieces to come.

## 6V6 Push-Pull Power Amp

The power amp for early Deluxe models is a bare-bones push-pull design with cathode bias. It has no grid-stopper resistors[23] or screen resistors. Grid stoppers prevent radiofrequency oscillation and affect the dynamics of overdrive. Screen resistors between the screen supply $V_{SS}$ and each screen also prevent radiofrequency oscillation. If their values are large enough, they prevent damage when the power tubes are pushed hard. They also affect harmonic distortion at full power.

---

[20] Tom Wheeler, **The Soul of Tone**, (Milwaukee: Hal Leonard, 2007), p. 155.
[21] John Teagle and John Sprung, **Fender Amps: The First Fifty Years**, (Milwaukee: Hal Leonard, 1995), p. 17. Tom Wheeler, **The Soul of Tone**, (Milwaukee: Hal Leonard, 2007), p. 216.
[22] Tom Wheeler, p.114.
[23] Richard Kuehnel, **Guitar Amplifier Electronics: Basic Theory**, (Seattle: Amp Books, 2018), pp. 70-72.

The Model 26's DC cathode voltage is somewhere in the neighborhood of 18V, making the DC grid-to-cathode bias equal to -18V, so the signal level at the grids needed for full power is 18V peak, which is +22.1 decibel-volts (dBV). (Quantifying audio signal levels using decibel volts is very convenient for guitar amplifier system design, as explained in the Appendix.)

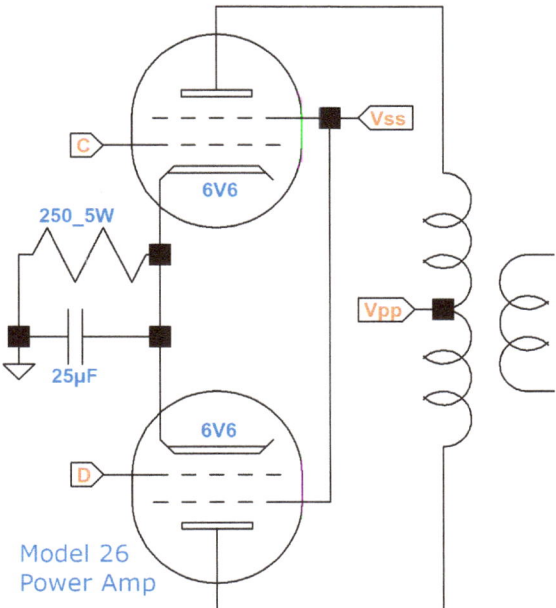

## 5Y3 Power Supply

The Model 26 power supply has a 5Y3 rectifier and a traditional *graded filter*, a series of individual AC ripple filters arranged so that the plate supply voltages applied to different tubes undergo different amounts of filtering.[24]

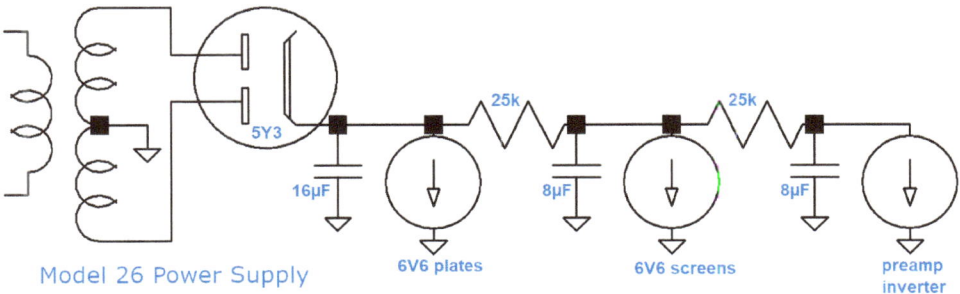

The 5Y3 vacuum tube rectifier has a relatively high plate resistance[25] that creates a generous amount of voltage sag when the power amp demands a lot of current. From the guitarist's perspective, the 5Y3 "provides lots of compression-like squash when the amp is cranked up and played hard."[26]

---

[24] Frederick Terman, **Radio Engineer's Handbook**, (New York: McGraw-Hill, 1943), p. 610.
[25] Richard Kuehnel, **Guitar Amplifier Electronics: Basic Theory**, (Seattle: Amp Books, 2018), pp. 40-43.
[26] Dave Hunter, **Guitar Amplifier Handbook**, Updated and Expanded Edition, (Milwaukee: Backbeat Books, 2015), p. 81.

The RC filters provide hum suppression and plate circuit decoupling.[27] Sufficient upstream AC ripple filtering is required to ensure that ripple levels for a stage's plate or screen supply are reasonably low compared to the audio signal levels that the stage is designed to handle. Plate circuit decoupling is needed to prevent multiple stages from breaking into low-frequency oscillation, a phenomenon known as *motorboating* because it sounds like the putt-putt-putt of an old-timer motorboat. Conventional wisdom dictates that a ripple filter be placed in the plate supply rail between every two inverting amplifier stages.

## 6N7 Paraphase Phase Inverter

The Model 26 second stage is a paraphase phase inverter implemented with a 6N7, which has two triodes sharing a common cathode, as shown by the pinout on the next page.

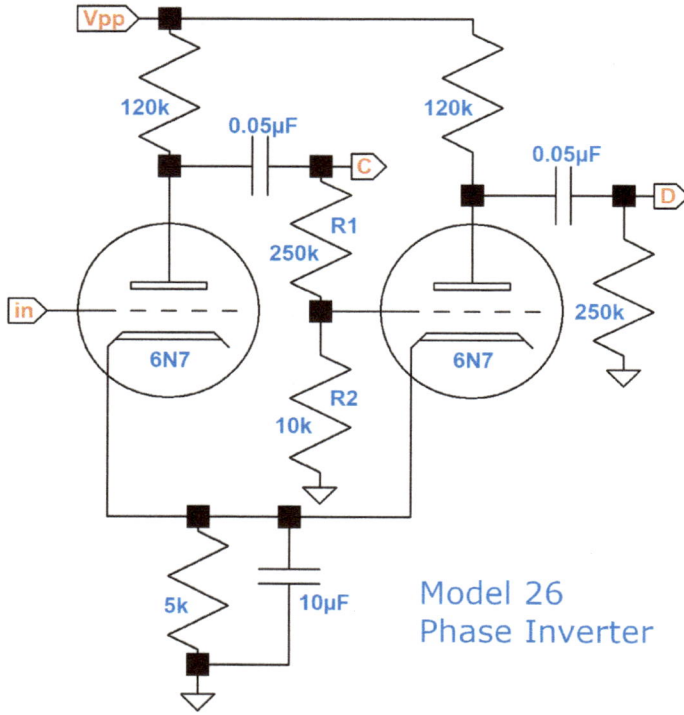

Model 26 Phase Inverter

"Paraphase makes sense when the only twin-triode is a common-cathode type. While the 6SL7 was available from 1941, I suspect it was priced higher and in shorter supply than the single cathode types. The

---

[27] Richard Kuehnel, **Guitar Amplifier Electronics: Basic Theory**, (Seattle: Amp Books, 2019), pp. 160-161.

6N7 was a near-miss in the 1930s market and may have been abundant by the 1940s."[28] –Paul Reid

The paraphase is Fender's inverter of choice until the company's narrow panel models shift to a split-load design. Afterwards Fender amps gradually move to the long-tailed pair and the paraphase disappears entirely from Leo's repertoire. The sonic characteristics of Fender's early amps are often attributed to their paraphase design.

> "[The paraphase] is an outdated PI topology that is prone to distortion in guitar amp circuits, and yields a spongy, compressed, mids-heavy sound."[29] –Dave Hunter

The metal-canned 6N7 is designed to withstand a maximum plate dissipation[30] of 5.5 watts for power amp applications, Class B push-pull operation in particular. Early versions of the Model 26 use a 6SN7.[31] Later versions have a 6SC7.[32] All of these are octal (8-pin) tubes.

**Pinout**    **Typical Small-Signal Parameters**

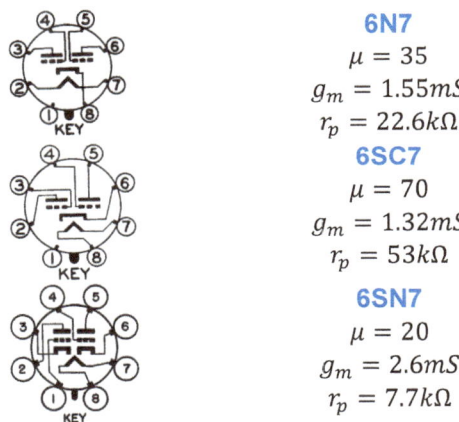

**6N7**
$\mu = 35$
$g_m = 1.55 mS$
$r_p = 22.6 k\Omega$

**6SC7**
$\mu = 70$
$g_m = 1.32 mS$
$r_p = 53 k\Omega$

**6SN7**
$\mu = 20$
$g_m = 2.6 mS$
$r_p = 7.7 k\Omega$

Outputs C and D drive the grids of the 6V6 push-pull power amp. As required for a paraphase inverter, resistors R1 and R2 form a voltage divider whose attenuation approximately matches the gain of the driving and driven stages. The amplifier-attenuator-amplifier combination creates two signals of opposite phase with approximately the same signal amplitude. The first triode amplifies and inverts the signal to drive one 6V6

---

[28] Personal correspondence with Paul Reid, February 2021.
[29] Dave Hunter, **Guitar Amplifier Handbook**, Updated and Expanded Edition, (Milwaukee: Backbeat Books, 2015), p. 80.
[30] Richard Kuehnel, **Guitar Amplifier Electronics: Basic Theory**, (Seattle: Amp Books, 2018), p. 104.
[31] John Teagle and John Sprung, **Fender Amps: The First Fifty Years**, (Milwaukee: Hal Leonard, 1995), p. 42.
[32] John Teagle and John Sprung, p. 43.

grid. The voltage divider attenuates the signal back to its former level so that it can be amplified and inverted again.

The unloaded voltage gain for a 6N7 amplification factor of 35, a 6N7 plate resistance of 22.6kΩ, a plate load resistor value of 120kΩ, and a fully bypassed cathode resistor is

$$\frac{(35)(120k\Omega)}{120k\Omega + 22.6k\Omega} = 29.5 \ (29.4dB)$$

The phase inverter's output impedance[33] is equal to the 120kΩ plate load resistor in parallel with the 6N7's 22.6kΩ plate resistance:

$$\frac{1}{\frac{1}{120k\Omega} + \frac{1}{22.6k\Omega}} = 19k\Omega$$

The second stage drives an AC load of 250kΩ, the grid-leak resistor for the second 6V6 power tube. Theoretically, the first stage should also drive 250kΩ but for practical reasons it is 10kΩ more to implement the attenuator, effectively creating a 260kΩ grid leak for the first power tube. For computational purposes, let's use their average value: 255kΩ. This means the 19kΩ output impedance and 255kΩ AC load create a voltage divider with a "gain" of

$$\frac{255k\Omega}{255k\Omega + 19k\Omega} = 0.93 \ (-0.6dB)$$

The loaded gain is therefore $29.4dB - 0.6dB = 28.8dB$.

It takes +22.1dBV signals (18V peak) at the 6V6 grids to drive the power amp to full power. At the phase inverter grid, the signal level is

$$22.1dBV - 28.8dB = -6.7dBV \ (654mV \ peak)$$

The attenuator is a voltage divider with a series resistance and shunt resistance equal to 250kΩ and 10kΩ, respectively. Attenuator "gain" is therefore

$$\frac{10k\Omega}{250k\Omega + 10k\Omega} = 0.038 \ (-28.3dB)$$

This is half a decibel less attenuation than required to fully compensate for the gain of the first voltage amplifier.

---

[33] Richard Kuehnel, **Guitar Amplifier Electronics: Basic Theory**, (Seattle: Amp Books, 2018), pp. 67-68.

A SPICE AC analysis simulation[34] shows that the second output (blue trace) has slightly more gain than the first output (red trace). Phase inverter imbalance creates 2nd-harmonic distortion, which is not necessarily bad for a guitar amp. The slight amount of bass attenuation is due to the two 0.05µF coupling capacitors. For an output impedance of 19kΩ, a 255kΩ average grid-leak resistor value, and a 0.05µF coupling capacitor, the -3dB break frequency[35] is

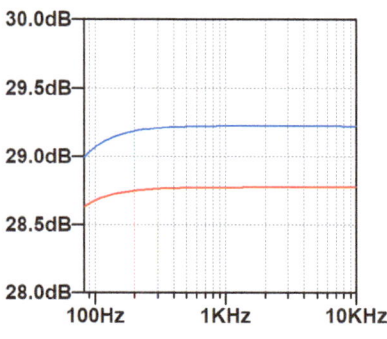

$$\frac{1}{2\pi(0.05\mu F)(19k\Omega + 255k\Omega)} = 12 Hz$$

This is well below guitar frequencies. At 82Hz, the lowest note on a guitar with standard tuning, the attenuation is less than a decibel. The signal for the second output passes through two coupling capacitors, which creates slightly more bass attenuation compared to the first output, as can be seen at the left end of the traces.

## Volume Controls

Between the preamp and phase inverter are volume controls that mix the instrument (A) and microphone (B) channels. Unlike contemporary designs, the wipers are used for the inputs and the tops of the potentiometers are the outputs. According to a Radio and Television Equipment Company magazine advertisement from 1947, this configuration implements "electronic mixing" of independent channels.

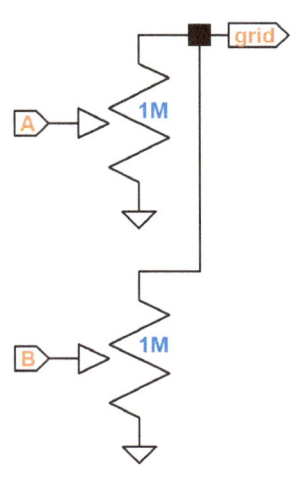

> "DELUXE AMPLIFIER FEATURES: 14 Watts Output. Heavy Duty 10" PM Speaker. Two instrument and one microphone input. Separate microphone and instrument volume controls. Electronic mixing of instrument and microphone inputs."[36]

A potentiometer's rotation corresponds to the division of resistance only if

---

[34] Richard Kuehnel, **Guitar Amplifier Electronics: Circuit Simulation**, (Seattle: Amp Books, 2019).
[35] Richard Kuehnel, **Guitar Amplifier Electronics: Basic Theory**, (Seattle: Amp Books, 2018), p. 22.
[36] Tom Wheeler, **The Soul of Tone**, (Milwaukee: Hal Leonard, 2007), pp. 116, 119.

it has a *linear taper*.[37] If the knob is set to 50-percent rotation, for example, then a potentiometer with a linear taper is set to 50-percent resistance – the wiper divides the total resistance into two equal halves.

Human perception of volume is logarithmic, so a *logarithmic taper* is commonly used for audio controls. Manufacturers specify a nonlinear taper in terms of percent resistance at 50-percent rotation or, alternatively, by providing graphs that depict percent resistance versus percent rotation. The data sheet for a CTS Electrocomponents Series 450G potentiometer conveniently provides both, as shown here.

| CTS Curve No. | % Resistance @ 50% Rotation |
|---|---|
| Std. D | 50% |
| Spl. D (EIA "S" Curve) | 50% |
| C | 5% |
| A | 10% |
| B | 15% |
| BD | 20% |
| H | 25% |
| J | 30% |

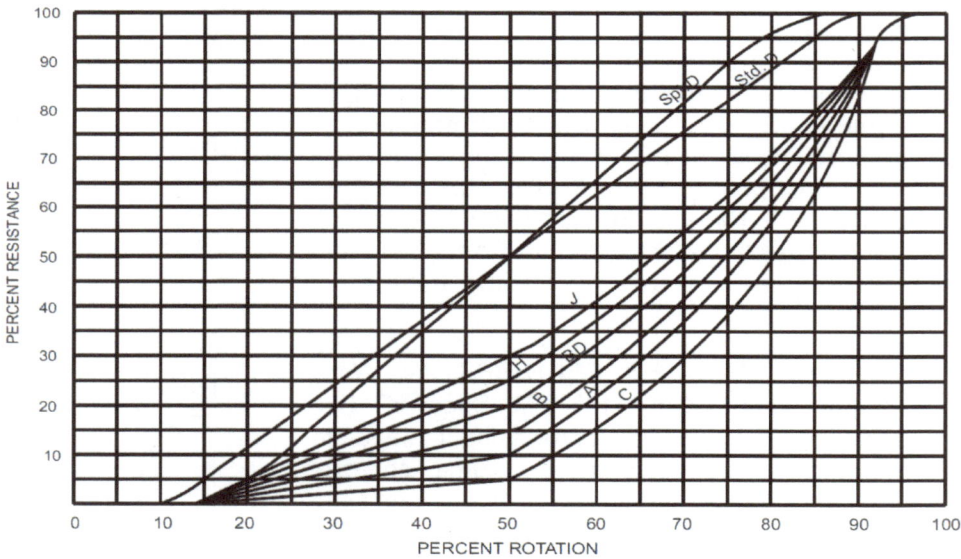

The top two curves "Std. D" and "Spl. D" are considered linear tapers. They have 50-percent resistance at 50-percent rotation. The other curves are nonlinear, with the bottom curve "C" being the most severe: at 50-percent

---

[37] Richard Kuehnel, **Guitar Amplifier Electronics: Basic Theory**, (Seattle: Amp Books, 2018), pp. 13-15.

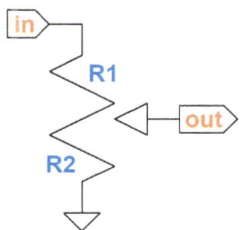

rotation the potentiometer is only at 5-percent resistance. When its knob position is set to 12 noon, there is only 5-percent resistance below the wiper, representing the shunt resistance R2, and 95-percent above, representing the series resistance R1.

For all of the amplifier circuits in this book we assume "audio" tapers have 10-percent resistance at 50-percent rotation, as depicted by CTS curve A.

The instrument input has a pair of 75kΩ resistors that provide 150kΩ isolation between guitar circuits when two instruments are used. With only one instrument connected, the resistors form a voltage divider that attenuates the signal by 6dB, as shown here.

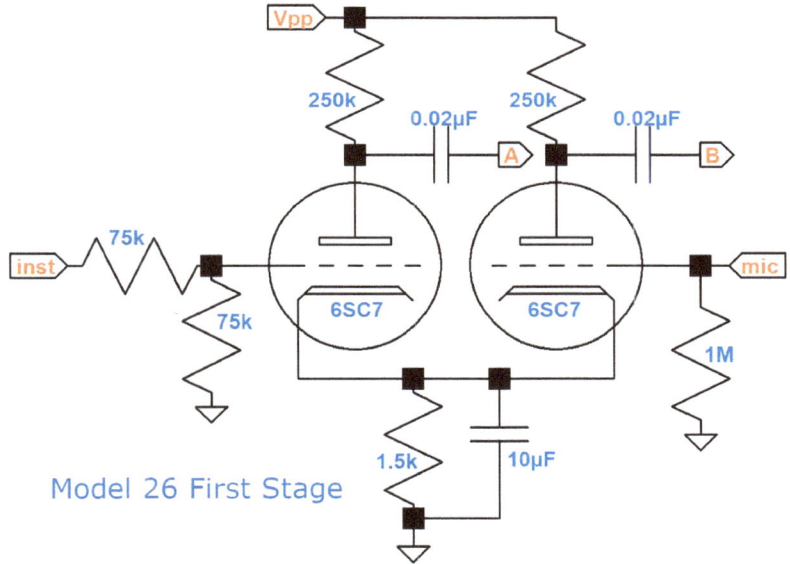

Model 26 First Stage

When only one guitar is plugged in, it makes sense to use the microphone input for more gain – attenuation prior to voltage amplification is not a desirable feature at the front end of a guitar amp.

Here is the Model 26 response from microphone input jack to the phase inverter grid, normalized to the response with the volume control set to maximum. The high-gain microphone channel volume control is set at 25-percent, 50-percent, 75-percent, and 100-percent rotation. The instrument channel's control is set to

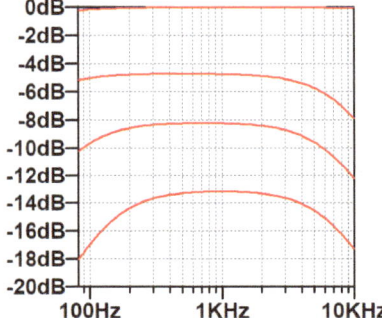

minimum.

At low frequencies and low control settings, the volume control represents a small shunt resistance to ground compared to the reactance[38] of the 0.02µF coupling capacitor, which increases bass attenuation. Similarly, at high frequencies there is more resistance in series with the 6N7 grid, which increases treble attenuation due to *Miller capacitance*[39] between the grid and the triode's other electrodes.

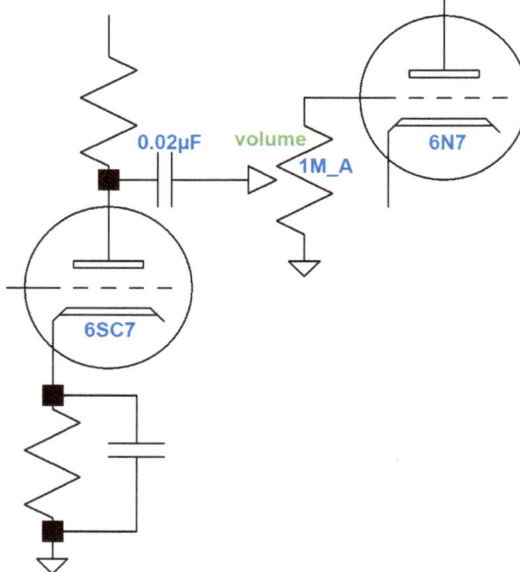

Most of the volume control's attenuation is caused by the AC load it places on the driving stage, which forms a voltage divider with the driving stage output impedance. To determine the *insertion loss*[40] when the active control is at 50-percent rotation (10-percent resistance) and the inactive channel's control is set to minimum, consider an equivalent circuit with fixed resistors.

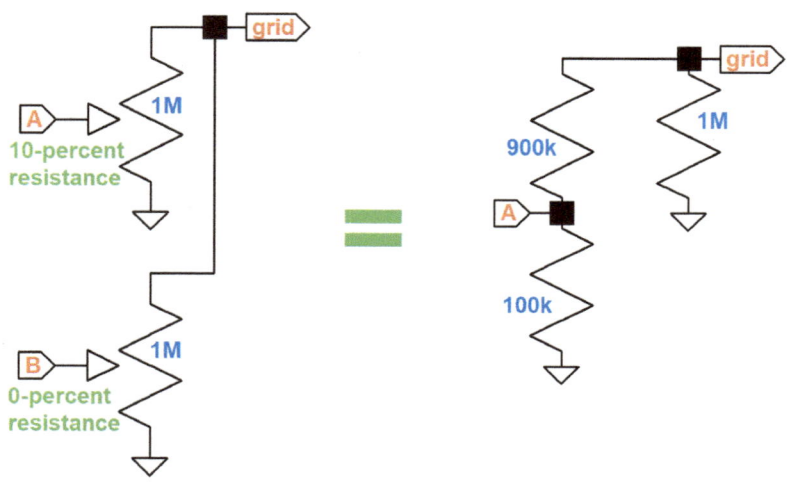

---

[38] Richard Kuehnel, **Guitar Amplifier Electronics: Basic Theory**, (Seattle: Amp Books, 2018), p. 19.
[39] **Basic Theory**. pp. 70-72.
[40] Richard Kuehnel, **Guitar Amplifier Electronics: Basic Theory**, (Seattle: Amp Books, 2018), p. 90.

**16 – CHAPTER 1**

The grid draws no current unless the phase inverter is in overdrive, so it does not contribute to the AC load of the first stage. The first-stage AC load is therefore equal to two resistances in parallel:

- 100kΩ (the resistance below the wiper) and
- 900kΩ (the resistance above the wiper) in series with the 1MΩ potentiometer for the other channel.

$$\frac{1}{\frac{1}{100k\Omega} + \frac{1}{900k\Omega + 1M\Omega}} = 95k\Omega$$

In the next section we determine that the output impedance of the driving stage is 44kΩ. The attenuation from the plate to the wiper is therefore

$$\frac{95k\Omega}{95k\Omega + 44k\Omega} = 0.68\ (-3.3dB)$$

From the wiper to the grid, the signal is attenuated by the voltage divider formed by the series resistance above the wiper (900kΩ) and the shunt resistance of the other control (1MΩ):

$$\frac{1M\Omega}{900k\Omega + 1M\Omega} = 0.53\ (-5.6dB)$$

The volume control's insertion loss at 50-percent rotation is therefore

$$3.3dB + 5.6dB = 8.9dB$$

With the microphone channel's control at maximum and the instrument control at minimum, the first stage output impedance drives a 500kΩ shunt resistance for a "gain" of

$$\frac{500k\Omega}{44k\Omega + 500k\Omega} = 0.92\ (-0.7dB)$$

## 6SC7 Preamp

The Model 26 first stage has a high-gain "microphone" input and two half-gain "instrument" inputs amplified by 6SC7 voltage amplifiers. As mentioned earlier, the instrument input jacks are wired so that if only one guitar is used, the second jack is shorted to ground, creating a voltage divider with the series and shunt resistances equal to 75kΩ, as shown on the next page. The voltage divider reduces the signal voltage by half, representing 6dB attenuation. The attenuation is not desirable – it is an unwanted but minor side effect of Fender's desire to isolate the guitar circuits from each other by placing 150kΩ between them.

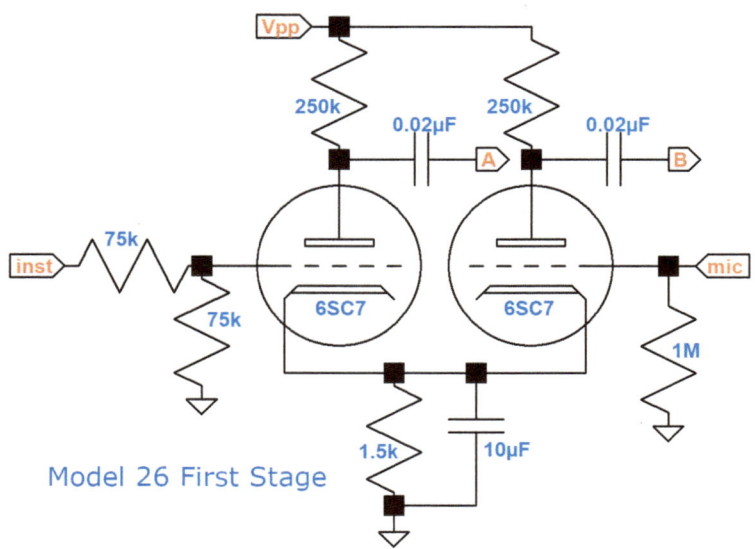

Model 26 First Stage

The 10µF cathode bypass capacitor acts as a short circuit for audio frequencies, making the triode circuits independent audio amplifiers. The 1.5kΩ cathode resistor carries the current of two triodes, so we can think of the circuit as two separate voltage amplifiers, each with a 3kΩ cathode resistor that is fully bypassed by a large capacitor.

Here is the equivalent circuit for the microphone input. A modern 12AX7 voltage amplifier with a 100kΩ plate load resistor and a cathode resistor that is fully bypassed has an unloaded gain of 36dB. The Model 26 preamp has a 250kΩ plate load resistor, which increases gain. The 6SC7 has a lower amplification factor compared to a 12AX7, however, and only a slightly lower plate resistance, so the net result is approximately the same voltage amplification.

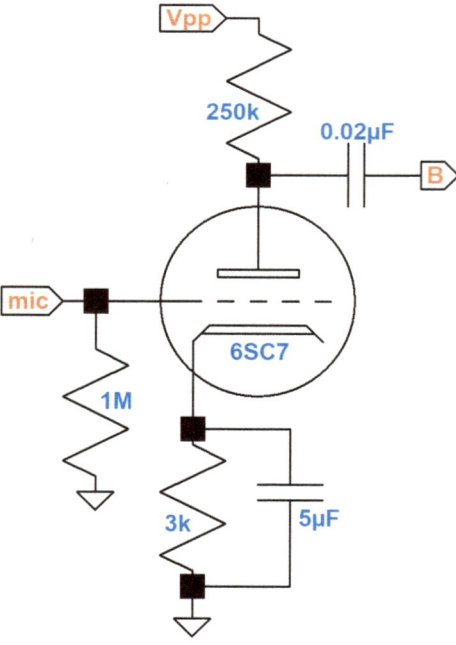

For the microphone input jack, the unloaded voltage gain for a plate load resistor value of 250kΩ, a typical 6SC7 amplification factor of 70, a typical 6SC7 plate resistance of 53kΩ, and a fully bypassed cathode resistor is

**18 – CHAPTER 1**

$$\frac{(70)(250k\Omega)}{250k\Omega + 53k\Omega} = 57.8 \ (35.2dB)$$

Gain is 6dB less for the instrument jack: 29.2dB. The output impedance, which we used in the last section to determine the insertion loss of the volume control circuit, is equal to the 250kΩ plate load resistor in parallel with the triode's plate resistance:

$$\frac{1}{\frac{1}{250k\Omega} + \frac{1}{53k\Omega}} = 44k\Omega$$

The signal level needed at the microphone input jack to drive the power amp to full power is equal to the signal level needed at the phase inverter input plus 8.9dB to account for volume control insertion loss minus the unloaded gain of the first stage:

$$-6.7dBV + 8.9dB - 35.2dB = -33dBV \ (32mV \ peak)$$

For an instrument input, 6dB more signal is required: -27dBV (64mV peak). The signal level at the jack required to drive the power amp to full power represents the input *sensitivity* of the amplifier for a specified set of control positions. Therefore, according to our calculations, the input sensitivity for the microphone jack, with its volume control set to 50-percent rotation and the instrument control set to minimum, is -33dBV (32mV peak).

Here is the grid-to-grid frequency response for the first stage and the volume control circuit, given the specified knob positions. Gain peaks at 26.4dB. Low-frequency attenuation from the coupling capacitor and high-frequency attenuation from Miller capacitance are clearly evident, although neither is severe.

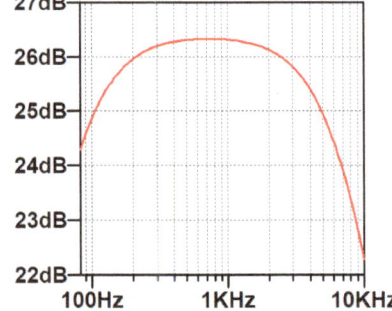

## Audio Signal Levels

The graph on the next page depicts audio signal levels in dBV at the input to each stage for a 1kHz signal driving the microphone input. The instrument volume control is at minimum and the power amp is operating at full power. The blue trace is for the microphone channel's volume control at 50-percent rotation (10-percent resistance). The orange trace is for the control at maximum rotation.

With the control at noon, the signal level at the jack required to drive the amp to full power, representing input sensitivity, is -33dBV (32mV peak).

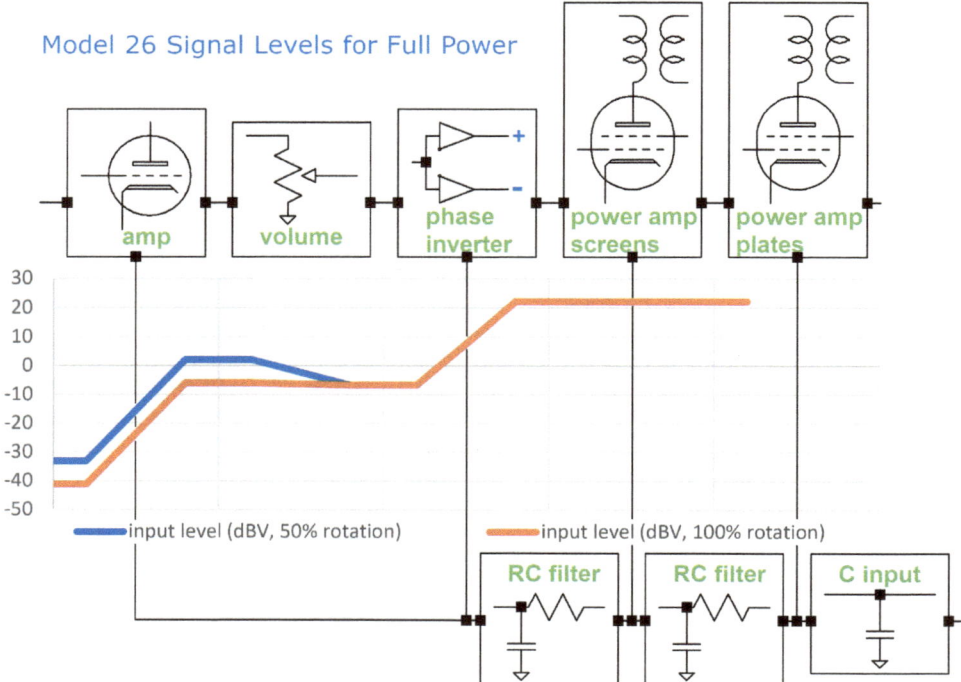

- The preamp boosts the signal by 35.2dB.
- The volume control circuit attenuates the signal by 8.9dB.
- The phase inverter boosts the signal by 28.8dB.

With the control at maximum, input sensitivity is -41.2dBV (12mV peak). For one guitar plugged into an instrument jack, there is 6dB less preamp gain. Sensitivities for 50-percent rotation and maximum rotation are then -27dBV (63mV peak) and -35.2dBV (25mV peak), respectively.

## Model 5A3 TV Front

As the TV-front amps appear, the Deluxe continues to occupy the middle of Fender's lineup, strategically positioned between the smaller 1x6 Champion and 1x8 Princeton and the bigger 2x10 Super and 1x15 Pro.[41]

---

[41] Tom Wheeler, **The Soul of Tone**, (Milwaukee: Hal Leonard, 2007), pp. 133-134.

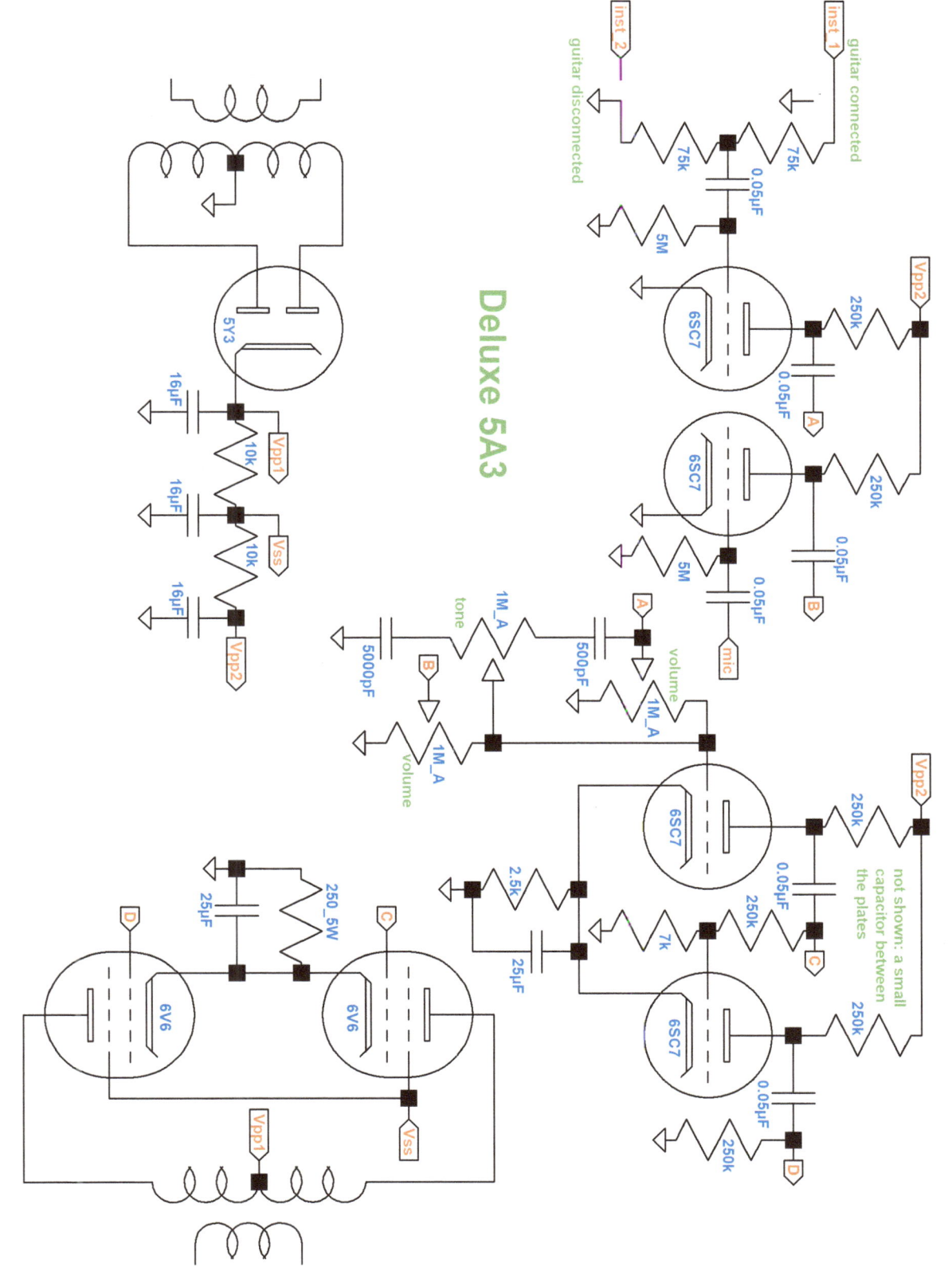

Of the five models, the 1x12 Deluxe is Fender's most popular, reflecting a sentiment that continues today.

> "The blackface version remains a definitive studio amp for the country player, and is great for anything from jazz to rock'n'roll besides. As for our TV-front Deluxe here, the combination of the octal 6SC7 preamp tubes, the paraphase inverter, lower voltages on the 6V6s, and a small paper-bobbin OT all make for a warm, chocolatey sound that flatters the traditional electric blues soloist like few other amps out there."[42] –Dave Hunter

## System Design Concept

In the early 1950s there is a growing desire among guitar players to tailor their sound using amplifier controls. The Model 26 has a crude treble-cut control between the power amp plates that affects both channels equally. The new Deluxe 5A3 improves the architecture by integrating a tone control with the preamp volume controls. The three controls affect the instrument and microphone channels uniquely and interactively, creating a broad palette of tones.

The new tone control concept becomes a key feature of the tweed Deluxe. A tone control circuit upstream of the power amp, however, unavoidably adds insertion loss. Between the volume control input and output, the guitar signal level takes a steeper plunge of attenuation when the controls are set to less than maximum. To help compensate, Fender increases phase inverter gain by replacing the 6N7 with a 6SC7. This creates a steeper climb of voltage gain toward the power amp grids.

Fender's new phase inverter design is not confined to the 6V6 Deluxe – Fender's larger, 6L6-based amps get the same circuit.

Another feature of the new Deluxe is more power. The 5A3 is at the front end of Fender's burst of circuit refinements designed to coax more power from its existing models. This is accomplished primarily by increasing power amp screen voltage.

Before we crunch the numbers, here is an overview of circuit enhancements to be examined in the sections ahead.

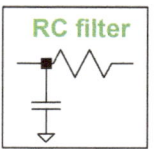 Ripple filter resistor values are halved and capacitor values are doubled to create higher plate and screen supply voltages without increasing AC ripple.

---

[42] Dave Hunter, **Guitar Amplifier Handbook**, Updated and Expanded Edition, (Milwaukee: Backbeat Books, 2015), p. 80.

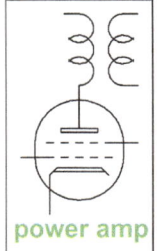
power amp

Higher 6V6 screen voltages translate to greater output power.

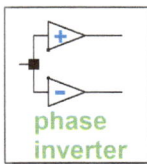
phase inverter

A second 6SC7 dual triode replaces the 6N7.

volume tone

A tone control is integrated with the preamp volume controls.

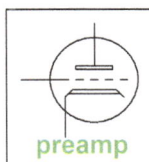
preamp

Grid-leak bias replaces cathode bias in the first-stage preamp.

## Power Amp and Power Supply Adjustments

The 5A3 uses the Model 26 power amp design but without a treble cut control between the 6V6 plates. The DC cathode voltage is in the neighborhood of 18V, making the grid-to-cathode bias equal to -18V and the signal level at the grids needed for full power equal to 18V peak (+22.1dBV).

As manufacturers make microfarads more affordable, the 5A3 power supply reduces the resistor values in the RC ripple

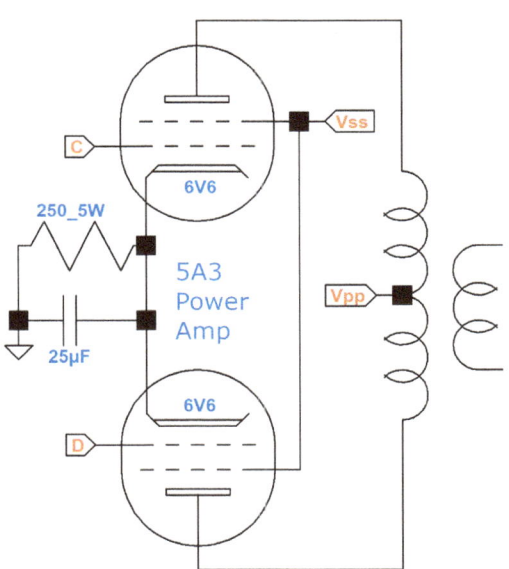

filters and compensates for the loss of ripple attenuation by increasing the capacitor values. With less RC resistance, the voltage drop across the resistors decreases, so DC supply voltages increase. For the 5A3 all three power supply capacitors are 16µF, not just the first.

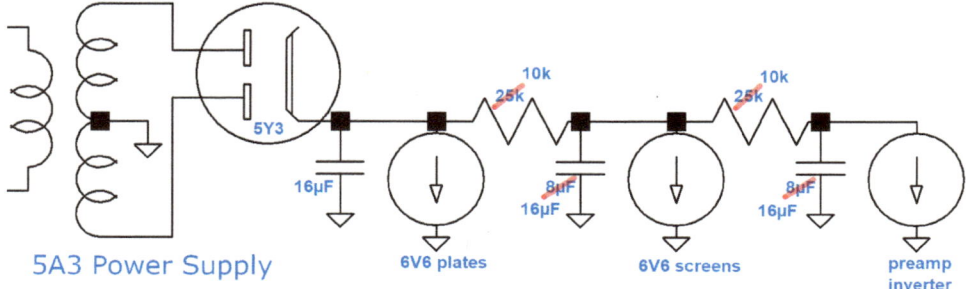

5A3 Power Supply         6V6 plates         6V6 screens         preamp inverter

## A New Phase Inverter Tube

The 5A3 uses a 6SC7 for the phase inverter. It also has a small capacitor between the plates (not shown) to reduce the chance of high-frequency oscillation.

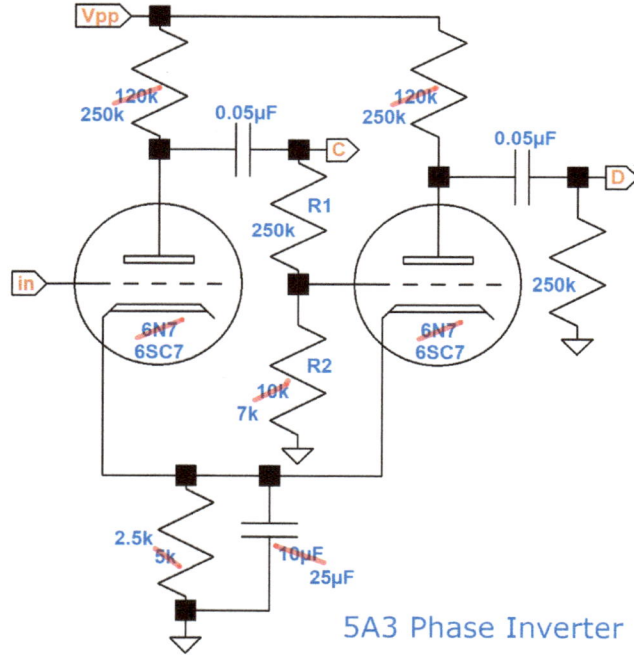

5A3 Phase Inverter

A paraphase phase inverter is comprised of two audio voltage amplifiers, making the 6SC7 a good choice. According to GE data sheets, it is a "high-mu twin triode designed primarily for audio frequency voltage amplification. Featuring a special shielding arrangement to reduce hum, the tube is well suited for low-level audio amplifier service." The 6N7, on the other hand, is "designed for use as a class B power amplifier in the output stage."

Fender's tube substitution doubles the amplification factor. Moreover, the 6SC7 needs only 300mA at 6.3V for its heater (the same as a 12AX7), compared to 800mA for the 6N7.

The Model 26 phase inverter has a loaded gain of 28.8dB. Insertion loss for its volume control circuit is 8.9dB with the active channel's control at noon. Net gain for the two together is 19.9dB.

For a 6SC7 amplification factor of 70, a 6SC7 plate resistance of 53kΩ, a plate load resistor value of 250kΩ, and a fully bypassed cathode resistor, 5A3 design changes increase the unloaded voltage gain by almost 6dB:

$$\frac{(70)(250k\Omega)}{250k\Omega + 53k\Omega} = 57.8 \; (35.2dB)$$

The output impedance is equal to the 250kΩ plate load resistor in parallel with the 6SC7's 53kΩ plate resistance:

$$\frac{1}{\frac{1}{250k\Omega} + \frac{1}{53k\Omega}} = 44k\Omega$$

The phase inverter's second triode drives an AC load of 250kΩ, the grid-leak resistor for the second 6V6 power tube. Theoretically, the first stage should also drive 250kΩ but for practical reasons it is 7kΩ more to implement the attenuator, effectively creating a 257kΩ grid leak for the first power tube. For computational purposes let's assume the average: 254kΩ. This means the output impedance and AC load create a voltage divider with a "gain" of

$$\frac{254k\Omega}{254k\Omega + 44k\Omega} = 0.85 \; (-1.4dB)$$

The loaded gain is therefore $35.2dB - 1.4dB = 33.8dB$. (For the Model 26 loaded gain is 5dB less: 28.8dB.)

It takes +22.1dBV signals (18V peak) at the 6V6 grids to drive the power amp to full power. At the phase inverter grid, the signal level is

$$22.1dBV - 33.8dB = -11.7dBV \; (368mV \; peak)$$

## Integrating Volume and Tone Controls

The 5A3 combines a tone control with the volume controls.

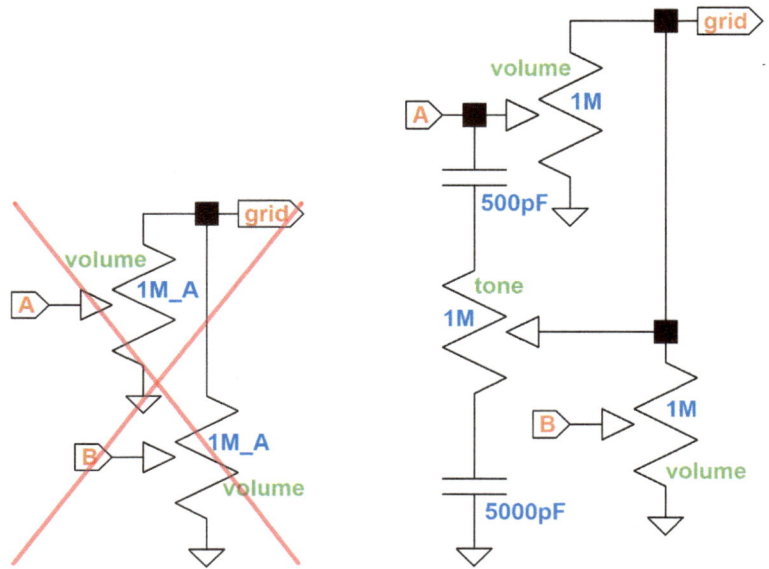

The tone control affects both the instrument and microphone channels, but in different ways. The controls are highly interactive – even the control for the unused channel has an effect on tone, which we will fully explore in Chapter 3. The interactivity gives the player the ability to get a variety of sounds out of the same instrument.

> "Most important in 1952 was the growing interest in the different 'sounds' available with 'electric' guitars, as opposed to 'electrified' ones. Having the ability to tailor one's sound through the controls of an amplifier, as well as to create a variety of sounds with the same instrument, was a fresh idea."[43]

For the instrument channel (node A), when the tone control is cranked up, it allows treble to bleed through the 500pF capacitor, bypassing the instrument volume control to create treble boost. If the volume control is at maximum, the input is connected directly to the grid and no further boost is possible. At low tone control settings, the 5000pF capacitor shorts high frequencies at the grid to ground, creating treble cut. The instrument channel volume control determines the range of tone control in a highly interactive way.

The output impedance of the 6SC7 preamp is 44kΩ, the same as for the Model 26. Here is a SPICE AC analysis simulation[44] of the volume and tone

---

[43] John Teagle and John Sprung, **Fender Amps: The First Fifty Years**, (Milwaukee: Hal Leonard, 1995), p. 27.
[44] Richard Kuehnel, **Guitar Amplifier Electronics: Circuit Simulation**, (Seattle: Amp Books, 2019).

control circuit when driven by that output impedance. The red traces are for the volume control set to 50-percent rotation (10-percent resistance). The blue traces are for the control set to maximum. The tone control is at minimum (lower traces), 50-percent rotation (10-percent resistance), and maximum (upper traces). The volume control for the microphone channel is set to minimum.

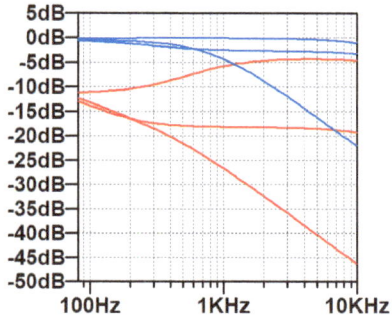

Except for a capacitor value in the 5B3, this volume and tone control circuit remains unchanged throughout the tweed era. Control interactivity will be further explored when we get to the iconic 5E3.

Here is the response of the volume and tone control circuit for a voltage source with a 44kΩ output impedance, the instrument volume and tone controls at 50-percent rotation (10-percent resistance), and the microphone volume control at minimum. At 1kHz the insertion loss is 19.1dB. When the volume and tone controls are set to maximum, the response is a flat -1dB.

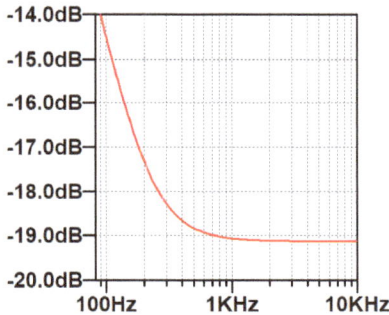

## Introducing Grid-Leak Bias

The first-stage preamp for the early tweeds represents a complete redesign of the Model 26 and a nostalgic return to the early days of vacuum tube technology.

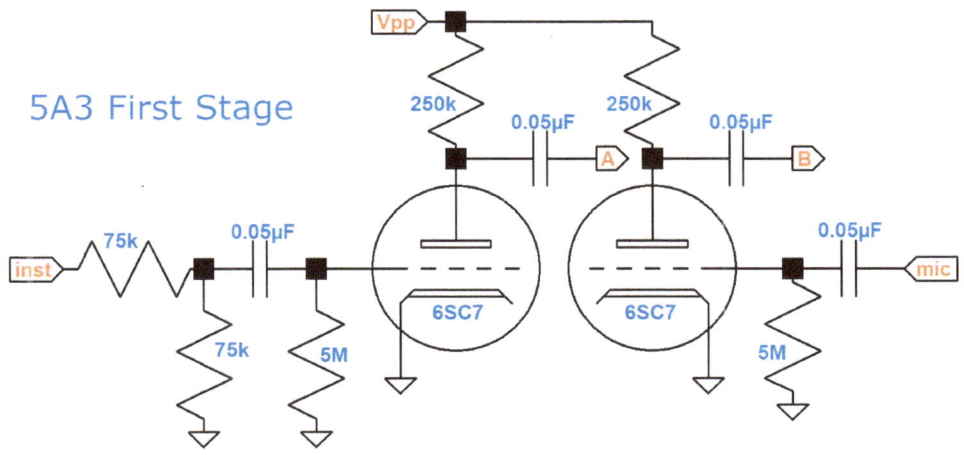

The 6SC7 and its 250kΩ plate load resistors remain. Gone, however, is the familiar cathode-bias of the Model 26. Instead, the 5A3 circuit uses *grid-leak bias*, the same biasing method found, for example, in a circa 1937 Rickenbacker amp.

The design change reduces the cost of the circuit by eliminating a large cathode bypass capacitor, which was relatively expensive at the time. The cathode is at *AC ground* in the Model 26 via the capacitor. In the 5A3 it is directly connected to ground.

We don't see grid-leak bias in modern amps because of unpredictable reactions to tube swaps and undesirable overdrive dynamics. Grid-leak bias is also susceptible to hum in low-level preamp stages.[45]

Grid-leak bias has three distinct features:

1. a grounded cathode,
2. a large grid-leak resistor value, and
3. a DC-blocking capacitor between the input and the grid.

For the more familiar cathode-biased first stage of the Model 26, the cathode resistor creates a negative grid-to-cathode voltage by making the cathode positive with respect to ground. If the cathode resistor value is large, it biases the tube cold by creating a more negative DC grid-to-cathode voltage. A small resistor value creates a warm (less negative) DC grid bias. With the 5A3's cathode connected directly to ground, the effective cathode resistor value is 0Ω and the DC grid bias is 0V, at least initially. This is hot bias in the extreme!

Making the cathode positive, like in a cathode-biased stage, cannot occur if it is grounded. Instead, the grid must somehow become negative with respect to ground. To accomplish this, the 5A3 relies on *bias excursion*.[46]

Bias excursion is caused by grid current between the grid and cathode within the tube. This current must come from somewhere and there are only two possible routes: via the extremely resistive 5MΩ grid leak or via the circuit at the front end of the coupling capacitor, which has a total resistance of tens of kiloohms at most. The latter is the most lucrative current path – positive grid current flows from the front end through the capacitor and then to the grid.

Unless the coupling capacitor suffers a catastrophic breakdown, electrons do not jump the gap between its plates. The only way the capacitor can pass current is to release electrons on one plate and accumulate electrons

---

[45] F. Langford-Smith, **Radiotron Designer's Handbook**, 4th ed., (Harrison: RCA, 1952), p. 489.
[46] Richard Kuehnel, **Guitar Amplifier Electronics: Basic Theory**, (Seattle: Amp Books, 2018), pp. 166-167.

on the other plate, causing it to charge. Negatively charged electrons emitted by the cathode are captured by the grid and passed on to a capacitor plate, making it negative with respect to the opposite plate connected to the guitar. This sets up a negative voltage at the top of the 5MΩ grid leak, creating a negative grid-to-ground voltage. When the amp is turned on, the DC grid voltage starts at 0V and then settles to an equilibrium value at which the capacitor stops charging. (Fender measures -0.9V for the Deluxe 5B3.)

Whether the cathode is directly connected to ground or merely at AC ground, there is no negative feedback from cathode degeneration[47] like there would be with an unbypassed cathode resistor. The 6SC7 voltage amplifier for the 5A3 therefore has the same unloaded gain and output impedance as for the Model 26 design: 35.2dB and 44kΩ, respectively. For the instrument jack, with the 75kΩ voltage divider between it and the 6SC7 grid, the unloaded gain from jack to plate is 6dB less: 29.2dB.

The signal level needed at the instrument jack to drive the power amp to full power is equal to the signal level required at the phase inverter input, plus 19.1dB to account for volume and tone control circuit insertion loss, minus the 29.2dB unloaded gain of the first stage:

$$-11.7 dBV + 19.1 dB - 29.2 dB = -21.8 dBV \ (115 mV \ peak)$$

We conclude that the input sensitivity for the instrument channel, with both of its controls at noon and the microphone channel's volume control set to minimum, is -21.8dBV (115mV peak).

Here is a SPICE AC analysis simulation of the frequency response from an instrument jack to the phase inverter grid when the microphone volume is at minimum and the other controls are at 50-percent rotation (10-percent resistance). Gain is increasing as the frequency decreases below 82Hz, due in part to the relatively large 0.05µF coupling capacitors. For this reason, the graph's lower limit is extended to 10Hz to show that eventually gain drops off for low frequencies, as it should.

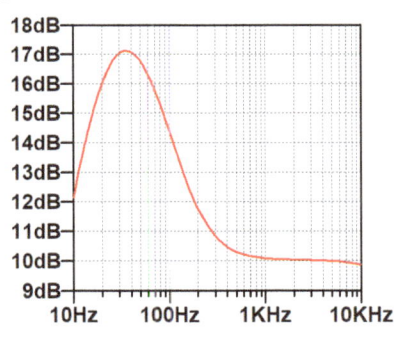

## Audio Signal Levels

Here are audio signal levels in dBV.

---

[47] Richard Kuehnel, **Guitar Amplifier Electronics: Basic Theory**, (Seattle: Amp Books, 2018), p. 61.

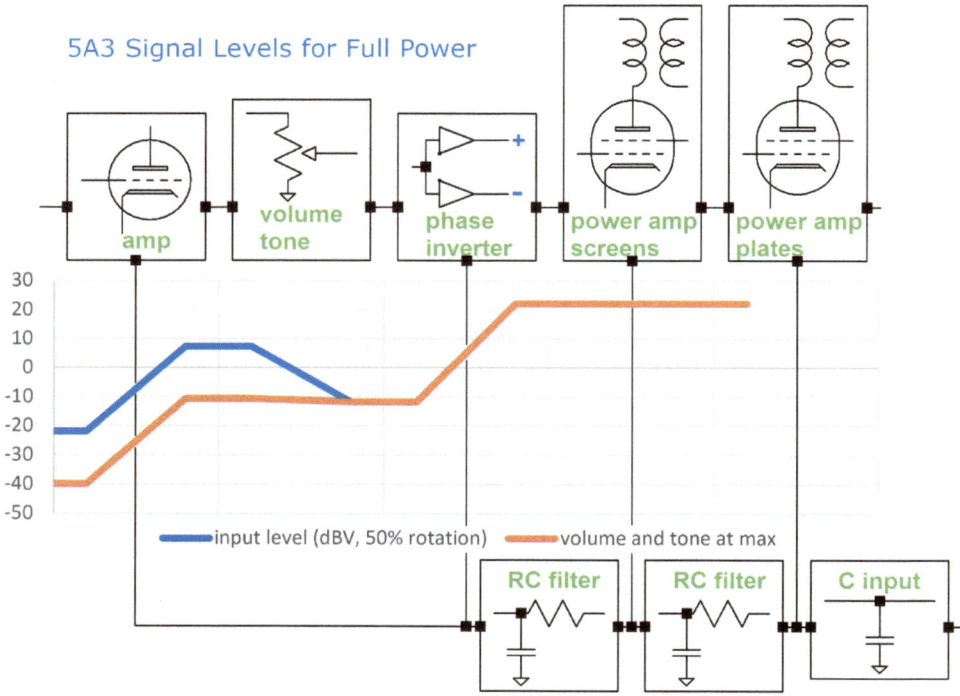

For the blue trace the instrument volume control is at 50-percent rotation, the tone control is at 50-percent rotation, the microphone volume control is at minimum, and the power amp is operating at full power. The input circuit and the preamp boost the signal by 29.2dB, the volume and tone control circuit attenuates the signal by 19.1dB, then the phase inverter boosts the signal to +22.1dBV (18V peak) at the 6V6 grids.

The 5A3's steeper plunge of tone control insertion loss, compared to the Model 26, is clearly evident. The steeper climb of phase inverter gain can also be seen, although it is less dramatic. Moreover, this time the guitar is plugged into the instrument jack, which reduces the first-stage gain by 6dB. (We are plugged into the instrument input for the 5A3 to use the new tone control as the designer intended.) The signal level at the jack for a 1kHz guitar signal with the instrument controls at 50-percent rotation is -21.8dBV (115mV peak), versus -33dBV (32mV peak) for the Model 26.

Because it is sufficient to drive the power amp to full power, -21.8dBV represents the amplifier's input sensitivity with the controls at 50-percent rotation. (A single-coil pickup has a maximum output of less than 500mV peak.) For the Model 26, sensitivity is -27dBV (63mV peak) because it lacks a tone control between the preamp and phase inverter.

When the controls are cranked to maximum, input sensitivity is -39.9dBV (14mV peak), which is well within single-coil capabilities, even for long sustains.

## Model 5B3 Wide Panel

"The wide panel 2x6V6GT Deluxe was a favorite of Scotty Moore, who reportedly used one on some of the most important recordings in the history of popular music."[48]

The wide-panel 5B3 retains the basic architecture of the 5A3 with no major circuit changes for the signal path. The new design offers only one minor upgrade.

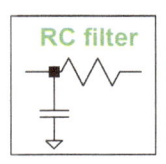

Ripple filter capacitor values increase to 20μF for slightly more ripple filtering.

"There's a real continuum in the circuits from tweed through blackface, and in some cases the changes were not nearly as drastic as the evolution of cabinet styles, colors, and other details might suggest."[49] – Matt Wilkens, Fender Musical Instruments

The DC voltages shown in the schematic on the next page are Fender's measurements, which claim an accuracy of plus or minus 20 percent. There is indeed a bit of slack in the numbers. According to Ohm's Law,[50] for example, the current through each preamp plate load resistor is

$$\frac{250V - 92V}{250k\Omega} = 0.63mA$$

Each phase inverter plate load resistor carries 0.4mA DC current. The total current through the 10kΩ RC ripple filter resistor upstream of the 250V plate supply is

$$2(0.63mA) + 2(0.4mA) = 2.06mA$$

This means the voltage drop across the resistor is

$$(2.06mA)(10k\Omega) = 20.6V$$

---

[48] Tom Wheeler, **The Soul of Tone**, (Milwaukee: Hal Leonard, 2007), p. 147.
[49] Tom Wheeler, p. 263.
[50] Richard Kuehnel, **Guitar Amplifier Electronics: Basic Theory**, (Seattle: Amp Books, 2018), pp. 9-12.

# Deluxe 5B3

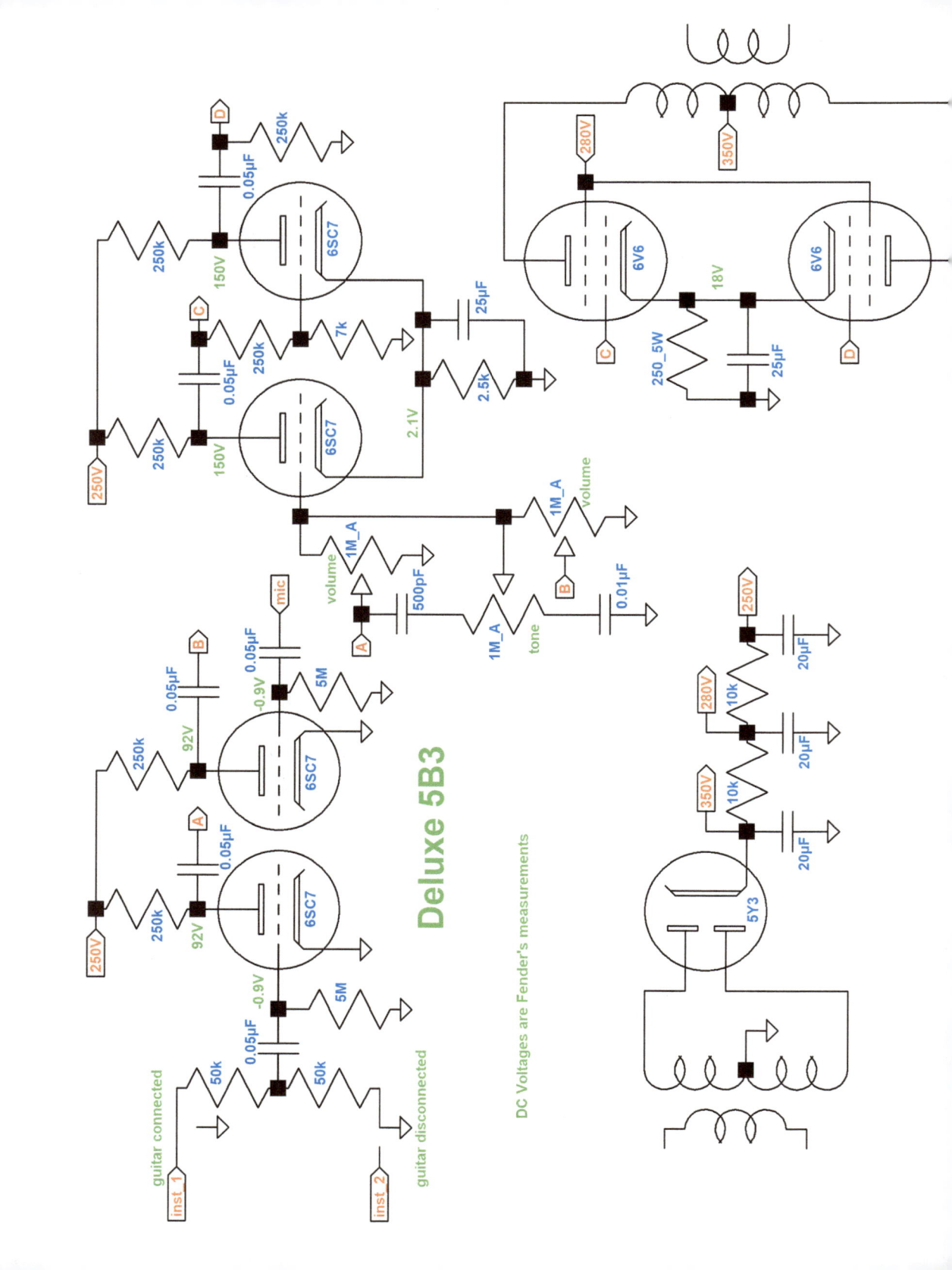

DC Voltages are Fender's measurements

Fender measures a 30V drop across the resistor. The biggest variants in this design are the DC operating points[51] for the first stage triodes. The results of grid-leak bias depend greatly on the particular tube occupying the socket, even for tubes from the same production run.

Right about the time that capacitor manufacturers shift from 4, 8, or 16 microfarads to 10, 20, or 40 microfarads, Fender upgrades its power supply.

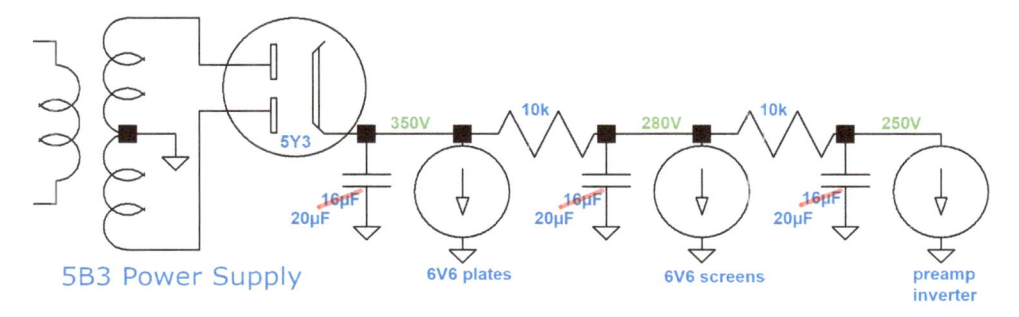

**5B3 Power Supply**

RC ripple filter attenuation increases by 2dB. This design change is short lived – the 5C3 reverts back to 16µF. We can speculate that with the next design iteration Fender gets a good deal on an old production run.[52]

The 5B3 reduces the resistor values at the instrument input jacks to 50kΩ. Attenuation is still 6dB when only one guitar is connected.

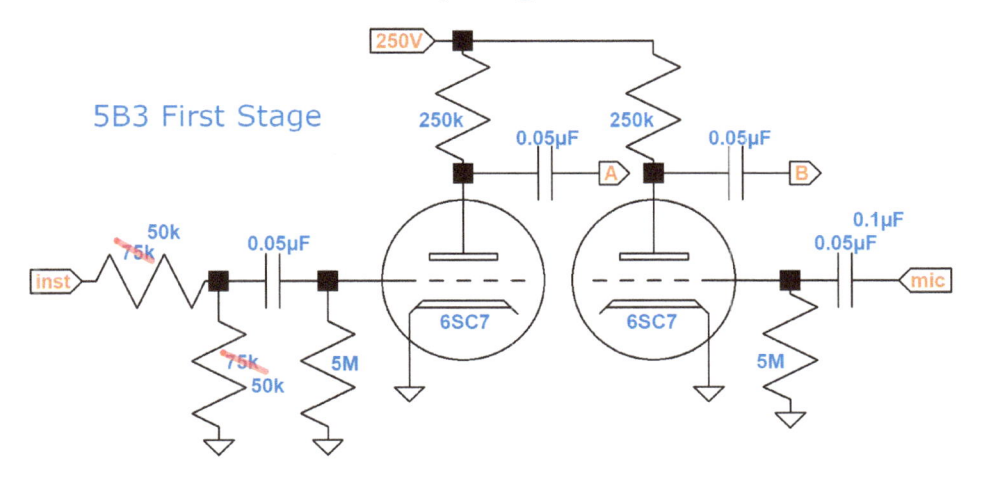

**5B3 First Stage**

The size of the tone control's shunt capacitor is doubled to 0.01µF. This shifts the lower limit of treble cut to lower frequencies.

---

[51] Richard Kuehnel, **Guitar Amplifier Electronics: Basic Theory**, (Seattle: Amp Books, 2018), p. 41.
[52] Personal correspondence with Paul Reid, February 2021.

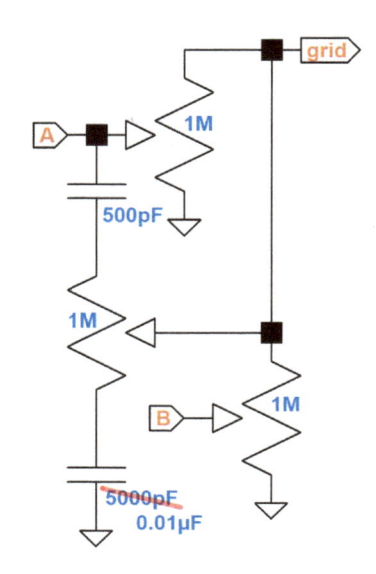

Below left is the SPICE AC analysis simulation we saw earlier of the 5A3 volume and tone control circuit when driven by a 44kΩ output impedance. Below right is the 5B3 with the larger capacitor.

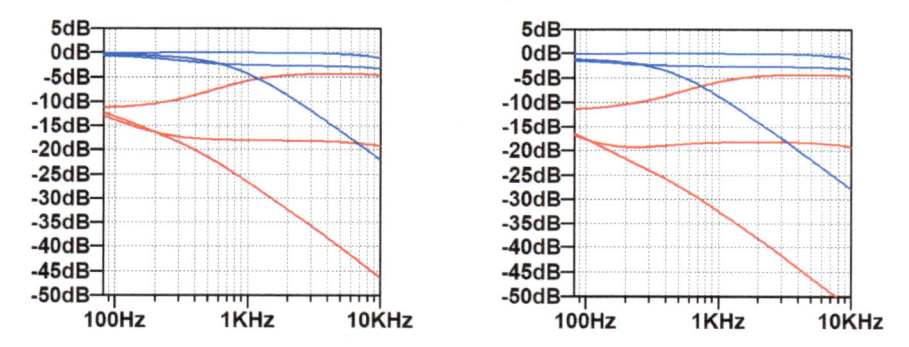

For these plots, the volume control is set to 50-percent rotation (10-percent resistance, red traces) and maximum (blue traces). The tone control is at minimum (lower traces), 50-percent rotation (10-percent resistance), and maximum (upper traces). The volume control for the microphone channel is set to minimum. The 5B3's larger capacitor value shifts the response to the left, but not in a dramatic way.

System audio performance for the 5B3 is nearly identical to the 5A3.

> "Armed with his Gibson ES-295 through a '52 Fender Deluxe, Scotty Moore, bassist Bill Black, and their bud Elvis, had no idea what a firestorm they were about to create with this early single."[53]

---

[53] "The 40 Most Important Guitar Solos in Rock," **Guitar Player**, April 6, 2017.

## Model 5C3 Wide Panel

The DC voltages in the 5C3 schematic on the next page are Fender's measurements. The DC currents are computed from these voltages using Ohm's Law.

### System Design Concept

During new product development, features are evaluated, amplifier performance objectives are set, and rough calculations are made to create a basic architecture.[54] Market factors are also considered, including the new product's price point in the context an entire product line. The Twin 5C8, for example, is a high-end product for demanding players who can afford it. The Deluxe 5C3, on the other hand, represents the middle of Fender's product line, where production costs are of increased importance. This at least partly explains why the Twin forges a different path than the Deluxe.

The high-end 5C8 adds separate bass and treble controls. To compensate for the increased insertion loss, the amp gets an extra voltage amplification stage courtesy of a single-triode octal: a 6J5. This adds another tube socket and associated support circuitry.

> "Upon its introduction at the 1952 NAMM show, the 5C8 Twin was the flagship of a new lineup. It was the first Fender with two 12" speakers and the first to carry independent Bass and Treble controls."[55]

The Deluxe 5C3, on the other hand, retains just a single tone control. A Deluxe with bass and treble controls does not appear until the blackface AA763.

The 5C3 introduces only a few changes to the 5B3.

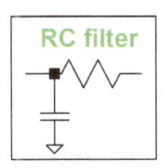

Ripple filter capacitor values revert back to 5A3 values: 16µF.

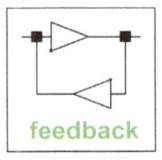

A small amount of negative feedback is introduced between the output transformer secondary and a power tube grid.

---

[54] Richard Kuehnel, **Fundamentals of Guitar Amplifier System Design**, (Seattle: Amp Books, 2019), p. 8.
[55] Dave Hunter, "The Fender 'Wide-Panel' Twin," **Vintage Guitar**, October 2015.

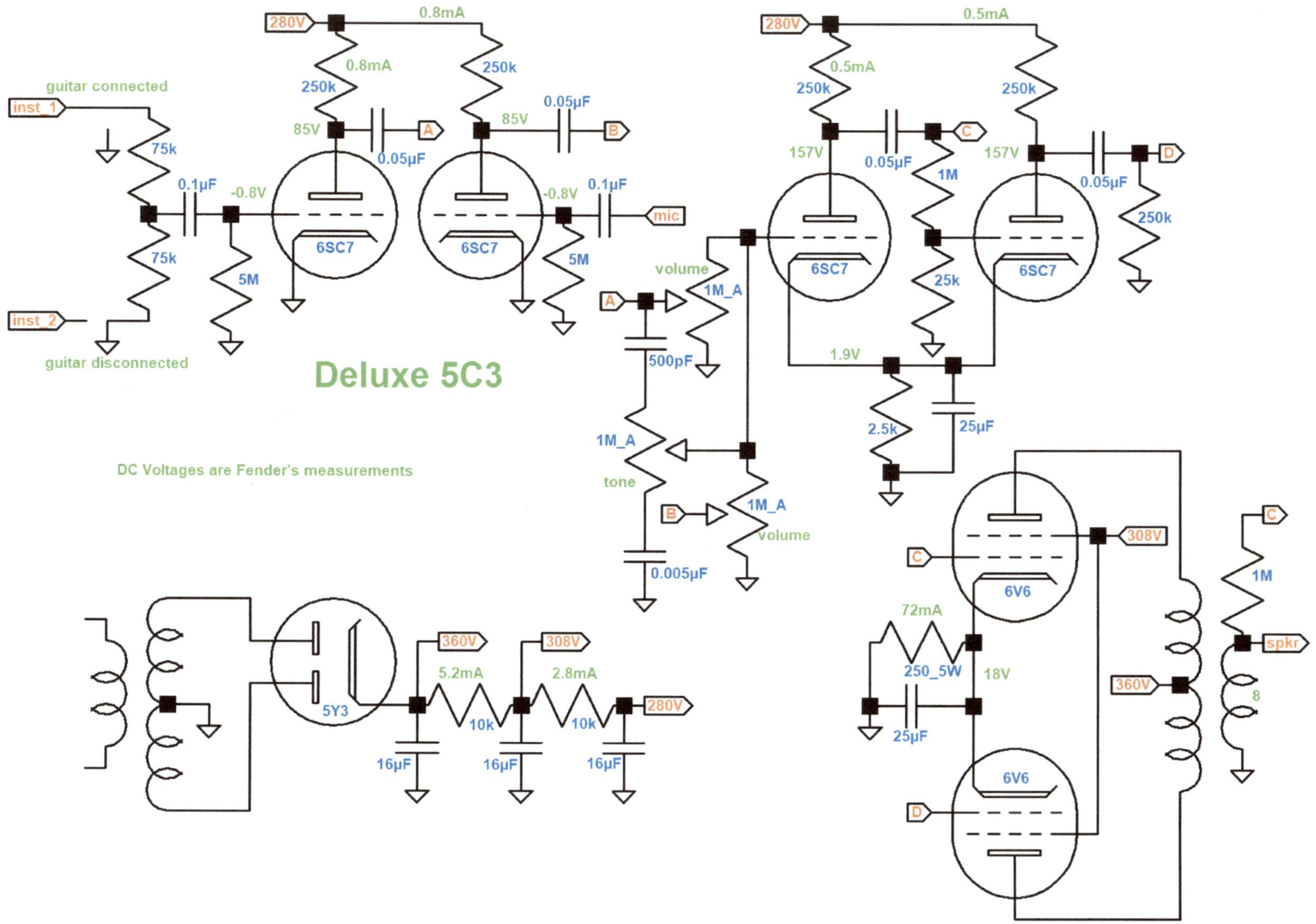

Deluxe 5C3
DC Voltages are Fender's measurements
guitar connected
guitar disconnected
inst_1
inst_2
75k
75k
0.1µF
-0.8V
5M
280V
0.8mA
250k
0.8mA
250k
85V
0.05µF
A
85V
0.05µF
B
6SC7
6SC7
-0.8V
0.1µF
mic
5M
volume
1M_A
A
500pF
1M_A
tone
B
1M_A
volume
0.005µF
280V
0.5mA
250k
0.5mA
250k
157V
0.05µF
C
1M
25k
157V
0.05µF
D
250k
6SC7
6SC7
1.9V
2.5k
25µF
308V
C
6V6
72mA
250_5W
18V
25µF
6V6
D
360V
C
1M
spkr
8
5Y3
360V
308V
5.2mA
2.8mA
10k
10k
280V
16µF
16µF
16µF

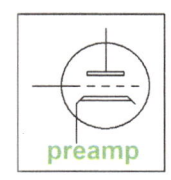

Resistor and capacitor values are tweaked.

## Preamp Tweaks

The 5C3 reverts back to the 5A3 value of 5000pF for the tone control circuit. The 5C3 also reverts back to the 5A3's 75kΩ resistors at the instrument jack. Attenuation is still 6dB when only one guitar is connected.

Coupling capacitor values at the input jacks are doubled to 0.1µF with negligible effect on the already flat bass response. The capacitor value does have an effect on overdrive, however. A large signal can cause grid current to flow, creating substantial distortion that diminishes as the capacitor charges.[56] If the capacitor value is too large, the transition can take too long.

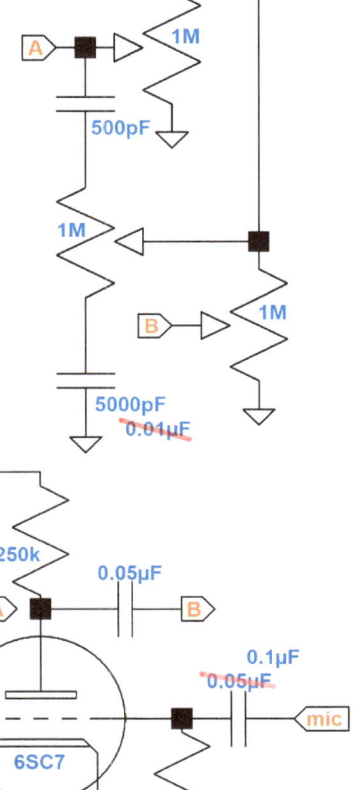

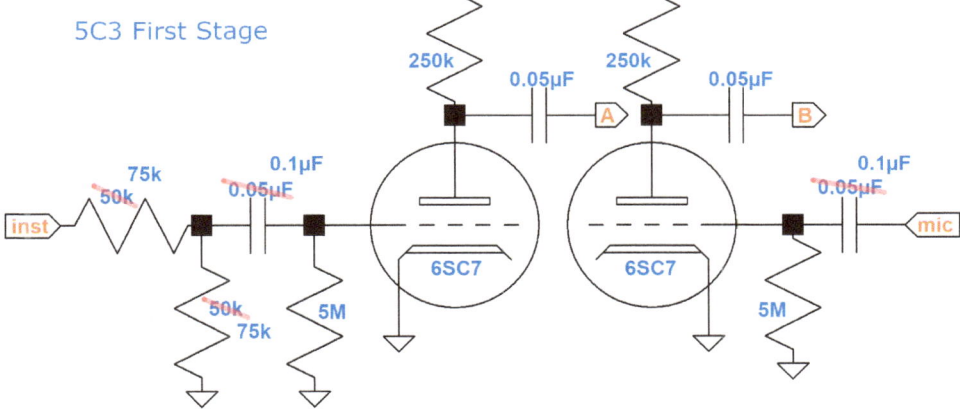

### 5C3 First Stage

System audio performance for the 5C3 is nearly identical to the 5B3.

---

[56] Arthur Evans, ed., **Designing with Field-Effect Transistors**, (New York: McGraw-Hill, 1981), pp. 116-117.

**HERITAGE – THE WOODIES AND EARLY TWEEDS – 37**

## Negative Feedback

Feedback is taken from the 8Ω tap via a 1MΩ feedback resistor. Parts values for the phase inverter's attenuator are adjusted to accommodate it.

The same feedback architecture with different parts values is used in the Twin 5C8, Super 5D4, and Pro 5D5, which use 6L6 power tubes. The effects of added feedback for the 5C3 are minimal for an 8Ω load (less than a decibel difference between open-loop and closed-loop gain[57]). Near the speaker's bass resonance frequency, on the other hand, forward gain can increase substantially, making feedback effects more pronounced.

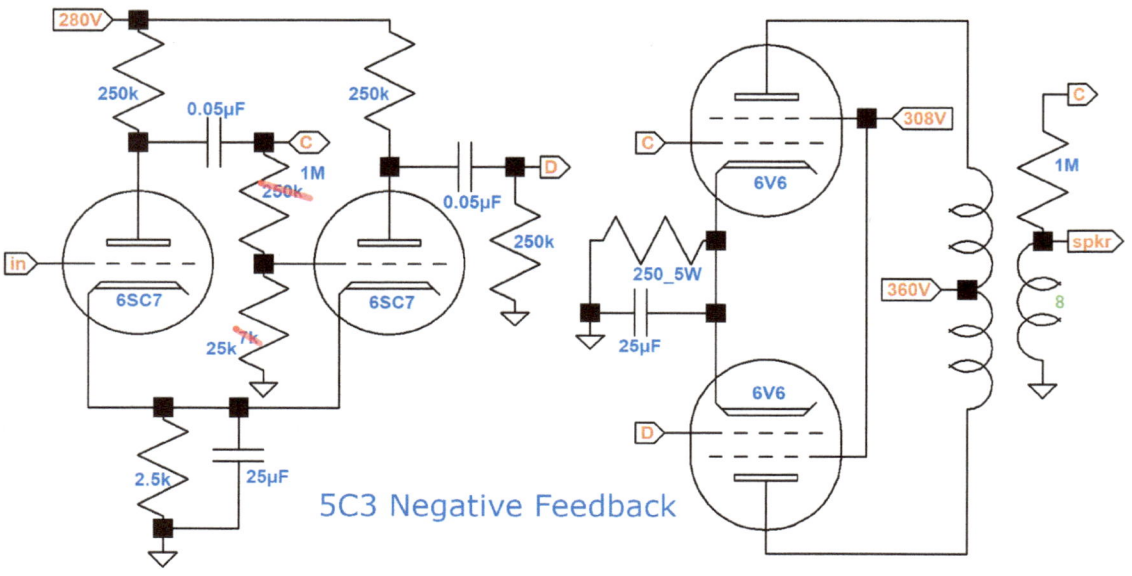

**5C3 Negative Feedback**

This feedback design is short lived. Just like the Pro 5E5, the Deluxe 5D3 reverts back to a design with no feedback. The Twin 5D8 and Super 5E4, on the other hand, advance the concept by increasing the amount of feedback and moving its insertion point to the phase inverter.

Negative feedback will be examined in detail starting with the 6G3, a Deluxe design that implements it in earnest for all guitar frequencies.

---

[57] Richard Kuehnel, **Guitar Amplifier Electronics: Basic Theory**, (Seattle: Amp Books, 2018), p. 146.

# Chapter 2: Legacy – The Wide Panel Deluxe 5D3

The Deluxe 5D3 represents the legacy design from which the iconic 5E3 is derived. This chapter takes a deep dive into the circuit and its performance to provide a context for the changes Fender introduces in the next design iteration.

## System Design Concept

The Deluxe hardly changes from the 5A3 to the 5C3. The 5D3, on the other hand, introduces major upgrades. Two modifications represent rollbacks of early "enhancements:"

1. a return to cathode bias for the preamp, as used in the Model 26, and
2. a removal of the modest amount of negative feedback introduced in the 5C3.

The first rollback gets rid of finicky grid-leak bias, a circa 1937 Rickenbacker-style design concept that eliminates the need for a large cathode bypass capacitor.

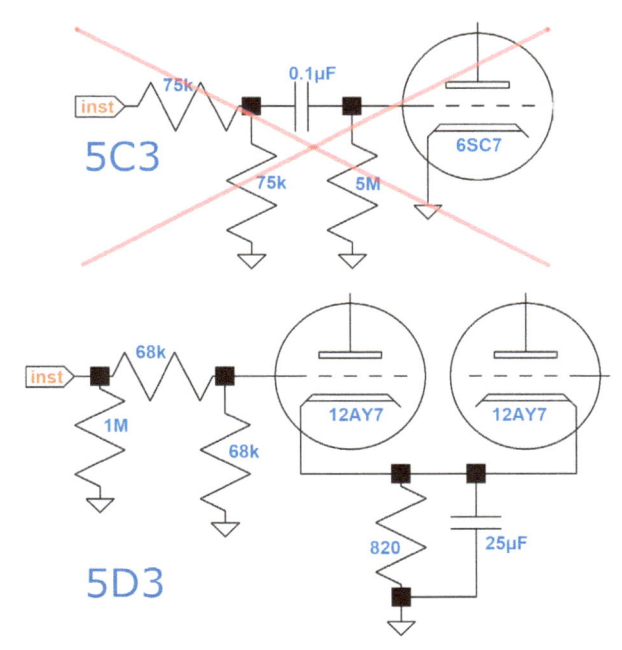

Cathode bias is more stable and predictable and remains the preamp biasing method of choice for modern guitar amps. Fender's mid-1950s return to cathode bias is not exclusive to the Deluxe – the Princeton 5C2, Super 5D4, Pro 5D5, and Twin 5D8 get the same rollback.

Negative feedback for the Bassman 5B6 extends from the output transformer secondary to the paraphase phase inverter, a concept that is later adapted to Fender's long-tailed-pair designs.

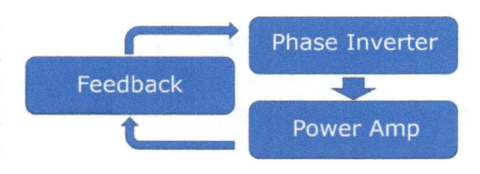

The Deluxe 5C3, Super 5D4, Pro 5D5, and Twin 5C8 leave the phase inverter out of the loop by routing a slight amount of feedback to the grid of a power tube. The implementation represents a timid and short-lived dabble into negative feedback. Subsequent Fender designs either advance the concept (Super 5E4 and Twin 5D8) by adapting the Bassman approach, or they revert back to a no-feedback design (Deluxe 5D3 and Pro 5E5).

Negative feedback reins in the wild whims of the loudspeaker, particularly for low frequencies, creating a "thumping and tight bass."[58] It also affects overdrive dynamics – the return to a no-feedback design for the Deluxe creates a power amp that runs wide open, with a more sustained, less snappy transition into overdrive.[59]

Fender's third 5D3 system design change represents true advancement: replacing the 8-pin octal preamp and phase inverter tubes with 9-pin miniatures. The Super 5C4 and Pro 5C5 are the first to transition to the new dual triodes. All other Fender models quickly follow.

> "We'd been using those larger [octal] preamp tubes, and we started developing circuits with the smaller ones. The older, larger ones were noisier and just plain troublesome. We got down to the 12AX7s, and they were much more compact and not as likely to have those microphonic problems."[60] –Don Randall

The 5D3 system design adjusts resistor values to accommodate the new tubes without substantially changing input sensitivity. Gain from the guitar input jack to the power tube grids is approximately the same.

Fender's schematic and layout for the 5D3 do not include DC voltage measurements, so the DC voltages and currents on the schematic on the next page are my estimates using 5Y3 and 6V6 data sheets, SPICE simulation, and Ohm's Law.

Here are circuit enhancements to be examined in the sections ahead.

---

[58] Dave Hunter, "Mezzabarba Custom Amplification M Zero Overdrive," **Guitar Player**, August 2018, p. 80.
[59] Richard Kuehnel, **Guitar Amplifier Electronics: Basic Theory**, (Seattle: Amp Books, 2018), p. 152.
[60] Tom Wheeler, **The Soul of Tone**, (Milwaukee: Hal Leonard, 2007), p. 222.

# Deluxe 5D3

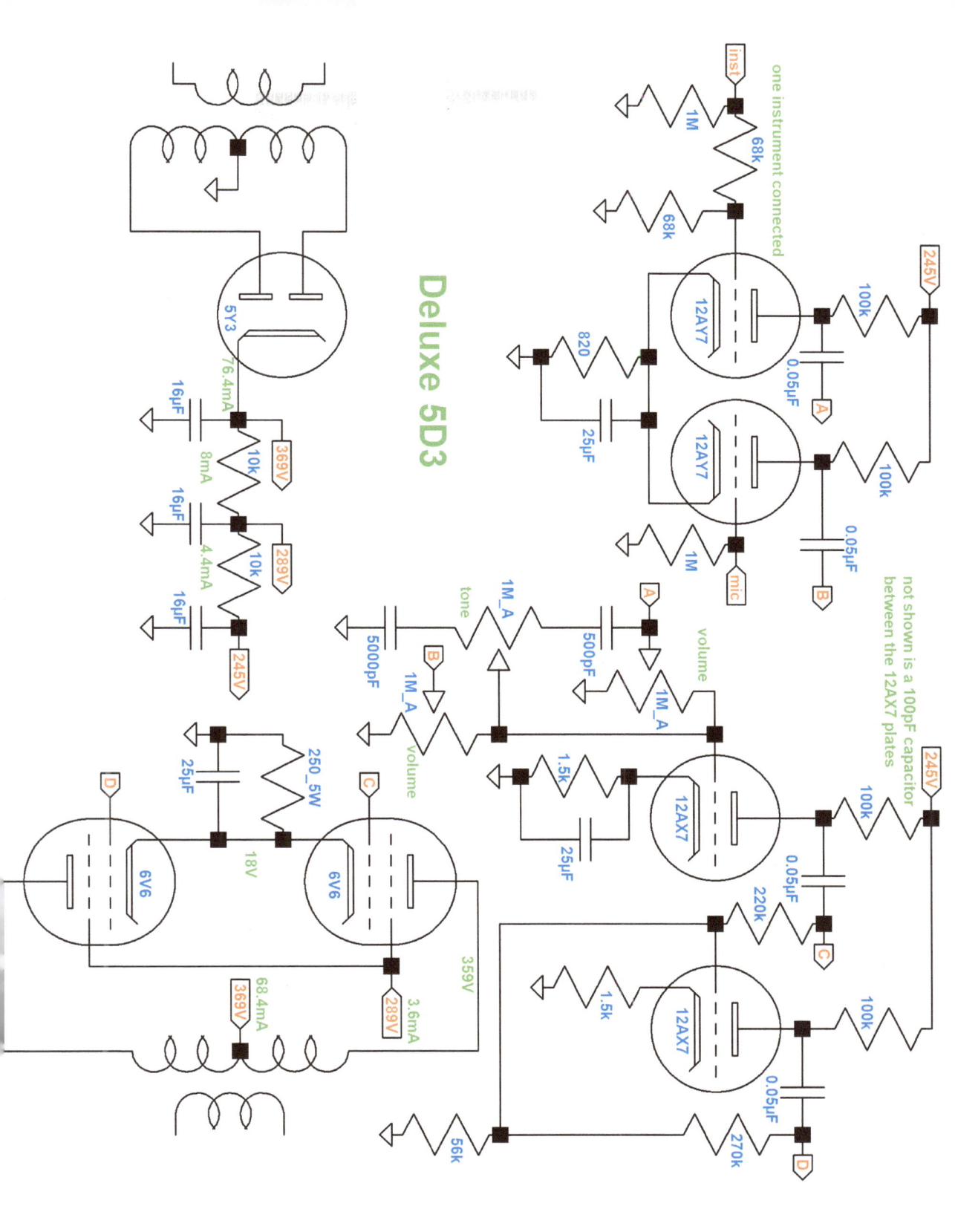

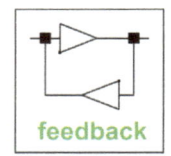

**feedback** — Negative feedback is eliminated. The power amp reverts back to the 5B3 design.

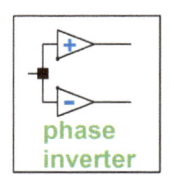

**phase inverter** — A high-mu 12AX7 replaces the 6SC7 for the phase inverter. The paraphase also becomes "self-balancing," a design shared by the Super 5D4, Pro 5D5, and Twin 5D8.

**preamp** — Cathode bias is reintroduced for the preamp. A medium-mu 12AY7 replaces the 6SC7.

## Power Amp Performance

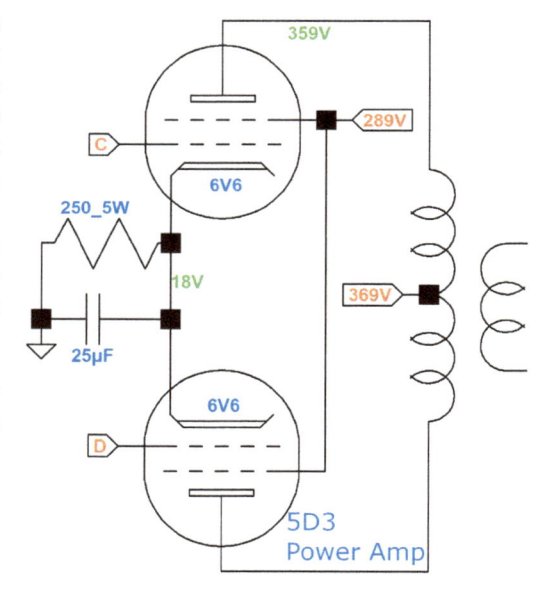

The 5D3 power amp retains the basic architecture of the 5C3 but without negative feedback. The screen-to-cathode voltage is 271V. A typical 6V6 datasheet only has graphs for plate and screen voltages up to 250VDC. On the next page are estimated *plate characteristics*,[61] which describe the relationship between plate current and plate-to-cathode voltage, for a 0V grid. Screen voltage ranges from 250V (blue curve) to 400V (yellow curve).

For the slope of the AC load line[62] we use one fourth the 8kΩ plate-to-plate transformer impedance: 2kΩ. The DC plate-to-cathode voltage is 341V. According to Ohm's Law, if the plate-to-cathode voltage swings to 0V, the plate current swings to

---

[61] Richard Kuehnel, **Guitar Amplifier Electronics: Basic Theory**, (Seattle: Amp Books, 2018), p. 38.
[62] **Basic Theory**, pp. 65-67.

$$\frac{341V}{2k\Omega} = 171mA$$

The red load line contains the endpoints 341V, 0mA and 0V, 171mA.

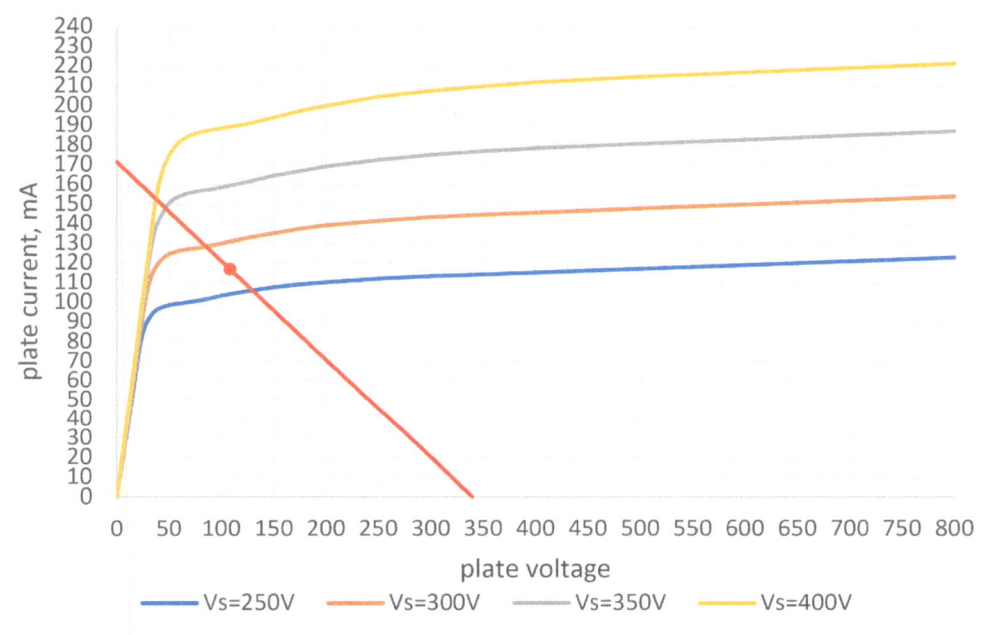

The load line intersects an estimated curve for 271V screen-to-cathode (red dot) at a plate voltage 105V and a screen voltage of 116mA for an output power of approximately

$$\left(\frac{1}{2}\right)(341V - 105V)(116mA) = 13.7W$$

The DC grid bias is -18V, so the power amp needs 18V peak (+22.1dBV) at the 6V6 grids, representing its input headroom, to achieve full power.

## Power Supply Ripple

The total power supply load current at idle is 76.4mA. For a DC load of 76.4mA at 369V, a 16µF capacitor input filter, and a ripple frequency of 120Hz, the ratio of the ripple voltage in RMS to DC voltage is[63]

$$\frac{\sqrt{2}(76.4mA)}{2\pi(120Hz)(369V)(16\mu F)} = 0.0243$$

---

[63] **Silicon Rectifier Handbook**, (Bloomington: Sarkes Tarzian, 1960), pp. 41, 42.

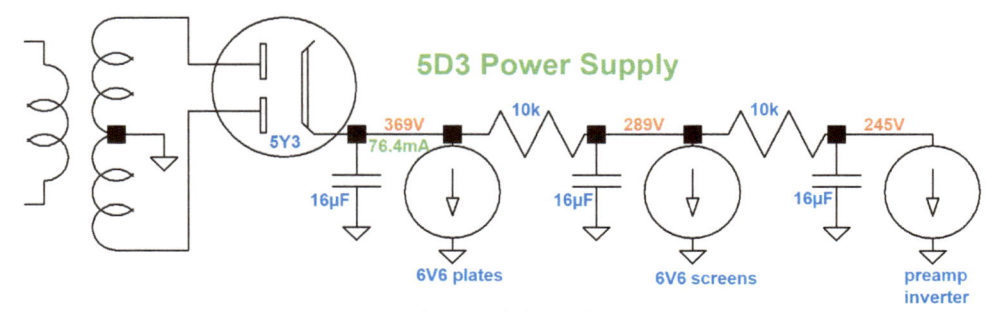

This means the ripple level is $(0.0243)(369V) = 9V\ RMS$, which is 13V peak (+19.3dBV).

Ripple attenuation for the first RC filter, with a 10kΩ resistor and 16μF capacitor, is

$$\frac{1}{\sqrt{[2\pi(120Hz)(10k\Omega)(16\mu F)]^2 + 1}} = 0.0083\ (-41.6dB)$$

AC ripple at the 6V6 screens is therefore

$$19.3dBV - 41.6dB = -22.3dBV\ (109mV\ peak)$$

The triode plate supply has an identical RC filter, so AC ripple for the triode supply is

$$-22.3dBV - 41.6dBV = -63.9dBV\ (903\mu V\ peak)$$

## Introducing a Self-Balancing Paraphase Phase Inverter

The 5D3 phase inverter shown on the next page uses a 12AX7 high-mu dual triode.

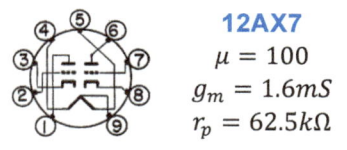

**12AX7**
$\mu = 100$
$g_m = 1.6mS$
$r_p = 62.5k\Omega$

Not shown in the 5D3 schematic is a 100pF capacitor between the plates to suppress radiofrequency oscillation.

The circuit is called a *floating paraphase*.[64] If R3 has slightly more resistance than R1, the voltage at the top of R2 *floats* close to ground potential. The circuit is self-balancing. If the amplitude at node C is greater than at node D, for example, it drives the voltage at the top of R2 in its

---

[64] F. Langford-Smith, **Radiotron Designer's Handbook**, 4th ed., (Harrison: RCA, 1952), pp. 524-525.

direction, which is fed to the grid of the second triode to increase the voltage swing at node D in the opposite direction.

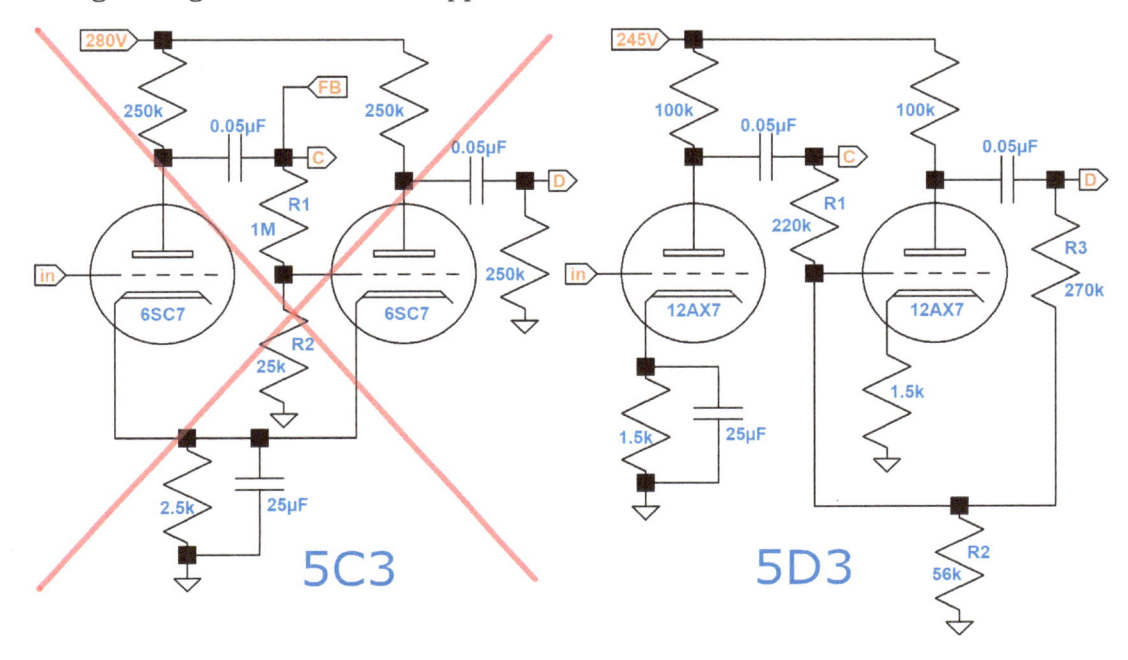

The first stage of the phase inverter is a standard 12AX7 voltage amplifier with a 100kΩ plate load resistor and a fully bypassed cathode resistor, a circuit that has become ubiquitous for guitar amplifiers.

> "In fact, as with his guitars, it's remarkable how often the term 'industry standard' comes up when discussing [Leo Fender's] amplifiers."[65] –Tom Wheeler

At DC the capacitors are open circuits and the grid of the first triode is at ground potential via a volume control. If the plate current is 0mA, then there is no current through the resistors and no voltage across them, so the plate-to-cathode voltage is equal to the 245V plate supply voltage. If, on the other hand, the plate-to-cathode

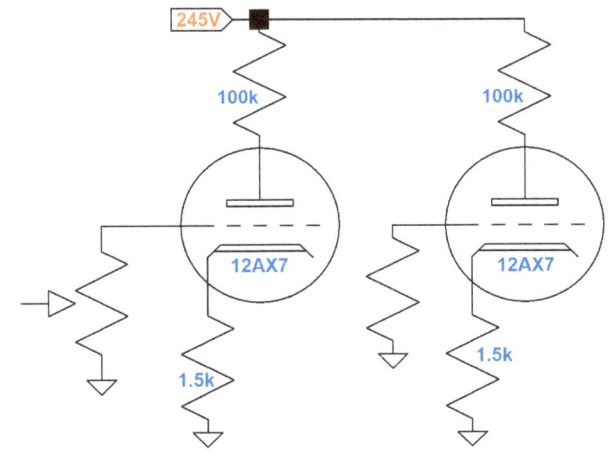

---

[65] Tom Wheeler, **The Soul of Tone**, (Milwaukee: Hal Leonard, 2007), p. 27.

voltage is 0V, then the plate and cathode are essentially short circuited. The total voltage across the two resistors in series is 245V. According to Ohm's Law, the current through the resistors is

$$\frac{245V}{100k\Omega + 1.5k\Omega} = 2.41 mA$$

These are the endpoints of the red DC load line[66] shown here.

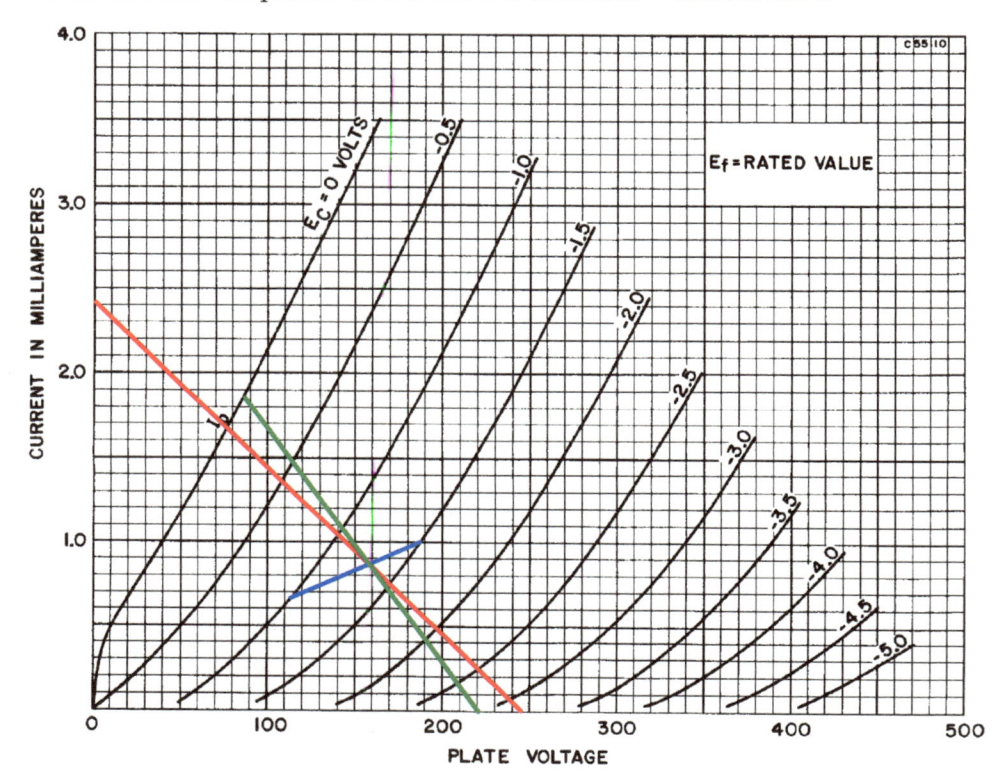

It is drawn on the 12AX7 plate characteristics, which have a separate curve for grid voltages ranging from 0V in the upper left to -5V in the lower right. Somewhere along the load line, which has a slope corresponding to the 101.5kΩ total resistance, is the DC operating point.

If the DC grid-to-cathode voltage is -1V or -1.5V, the cathode voltage is 1V or 1.5V, respectively. According to Ohm's Law, the values for plate current passing through the 1.5kΩ cathode resistor are

---

[66] Richard Kuehnel, **Guitar Amplifier Electronics: Basic Theory**, (Seattle: Amp Books, 2018), pp. 57-58.

$$\frac{1V}{1.5k\Omega} = 0.67mA$$

$$\frac{1.5V}{1.5k\Omega} = 1mA$$

These values are the endpoints for the blue line segment. It intersects the load line at the DC operating point: -1.3V grid-to-cathode, 159V plate-to-cathode, and 0.9mA plate current.

The top of resistor R2 is a floating ground, giving the first triode an AC load of 220kΩ. The slope of the AC load line is 100kΩ in parallel with the AC load:

$$\frac{1}{\frac{1}{100k\Omega} + \frac{1}{220k\Omega}} = 69k\Omega$$

According to Ohm's Law, a plate current swing from 0.9mA at the DC operating point to 0mA creates a voltage swing of

$$(0.9mA)(69k\Omega) = 62V$$

The plate voltage therefore swings from its DC value of 159V to a value of $159V + 62V = 221V$. The AC load line therefore connects the DC operating point to 221V, 0mA, as shown in green.

According to the AC load line, the grid can swing by up to 1.3V peak without severe distortion, so *input headroom* is 1.3V peak (-0.7dBV). If the circuit is perfectly balanced, the second triode's plate swings by the same amount as the first, so its headroom limits are the same.

The loaded gain for a 12AX7 amplification factor of 100, a 12AX7 plate resistance of 62.5kΩ, an AC plate load of 69kΩ, and a fully bypassed cathode resistor is

$$\frac{(100)(69k\Omega)}{69k\Omega + 62.5k\Omega} = 52.5 \ (34.4dB)$$

The power amp needs +22.1dBV at the 6V6 grids for full power. The corresponding signal level at the phase inverter input (the first triode grid) is

$22.1dBV - 34.4dB = -12.3dBV \ (343mV \ peak)$

Here is the frequency response from the first 12AX7 grid to the first 6V6 grid (red) and second 6V6 grid (blue). The relatively

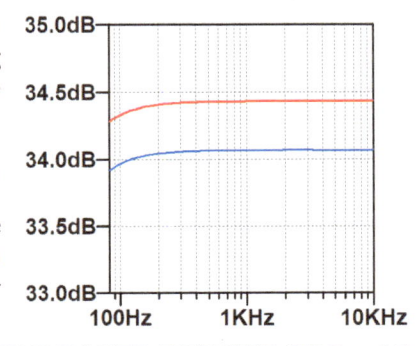

**LEGACY – THE WIDE PANEL DELUXE 5D3**

large 0.05μF coupling capacitors create a flat bass response, with only about 0.1dB attenuation at 82Hz.

## Volume and Tone Control Insertion Loss

The volume and tone control circuit is identical to the 5C3. The controls are highly interactive, a feature that we will explore when we get to the 5E3 in the next chapter.

With both knobs cranked to maximum, the AC load creates 0.5dB attenuation. With both knobs at 50-percent rotation (10-percent resistance), insertion loss is about 17.4dB at 1kHz, with less attenuation for bass and slightly more for treble, as shown here.

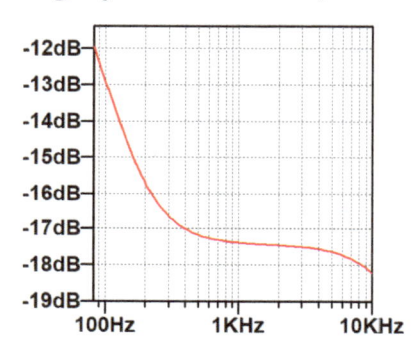

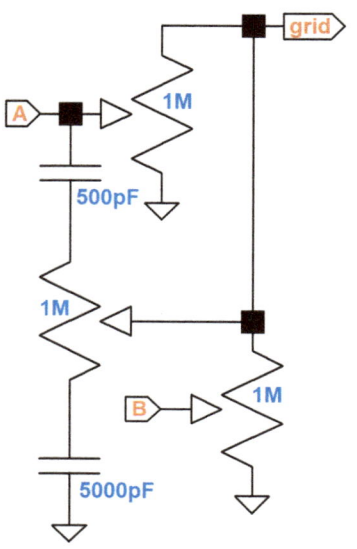

The plot assumes the volume control for the microphone channel is set to minimum. It also assumes the signal source has a 20kΩ output impedance, which corresponds to the 5D3's first stage, as we will determine in the next section. The unloaded signal level at the driving stage output needed for full power is

$$-12.3 dBV + 17.4 dB = +5.1 dBV \ (2.5V\ peak)$$

With the controls at maximum, the required signal level is 16.9dB less:

$$-12.3 dBV + 0.5 dB = -11.8 dBV \ (364mV\ peak)$$

## A New Preamp Design

The 5C3 is the last Deluxe to use grid-leak bias. Instead, the 5D3 implements cathode bias with a new triode: a 9-pin miniature 12AY7.

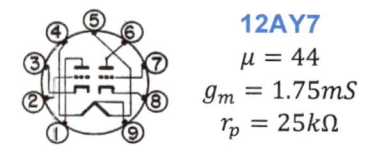

**12AY7**
$\mu = 44$
$g_m = 1.75 mS$
$r_p = 25 k\Omega$

## 5D3 First Stage

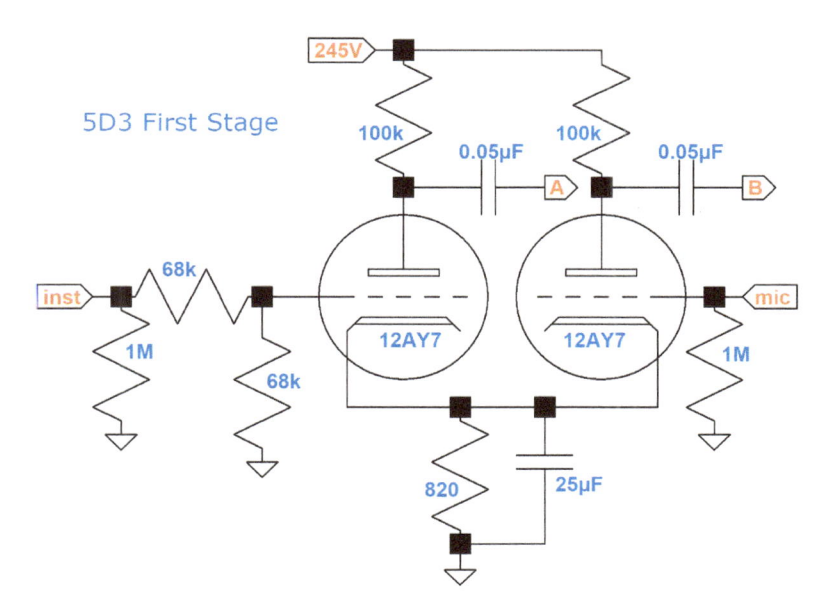

With a pair of 68kΩ resistors, the instrument input is still configured to create 6dB attenuation when only one guitar is used, so the voltage swing at the grid is half the swing at the jack. When two guitars are plugged in, the 1MΩ grid-leak resistor guarantees an electrical path from grid to ground – the guitar circuits do not serve as grid-leak resistors like they do for the Model 26.

For DC the capacitors are open circuits. The 820Ω cathode resistor carries the cathode current of two triodes, so to create the same cathode voltage with only one triode the resistor value needs to be doubled to 1.64kΩ. For the purposes of determining the DC operating point, this is the equivalent circuit for one triode.

If the plate current is 0mA, then the plate-to-cathode voltage is 245V. If, on the other hand, the plate-to-cathode voltage is zero, then there is 245V across a total resistance of $100k\Omega + 1.64k\Omega = 101.64k\Omega$. According to Ohm's Law, the plate current through the resistors is then

$$\frac{245V}{101.64k\Omega} = 2.41mA$$

The red DC load line therefore connects 245V to 2.41mA.

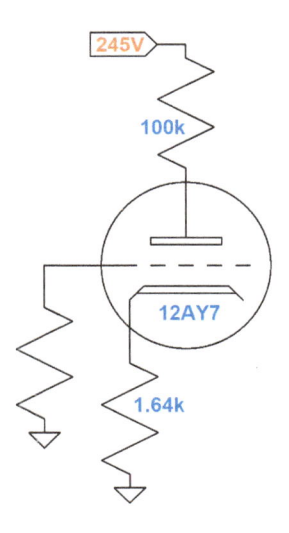

**LEGACY – THE WIDE PANEL DELUXE 5D3**

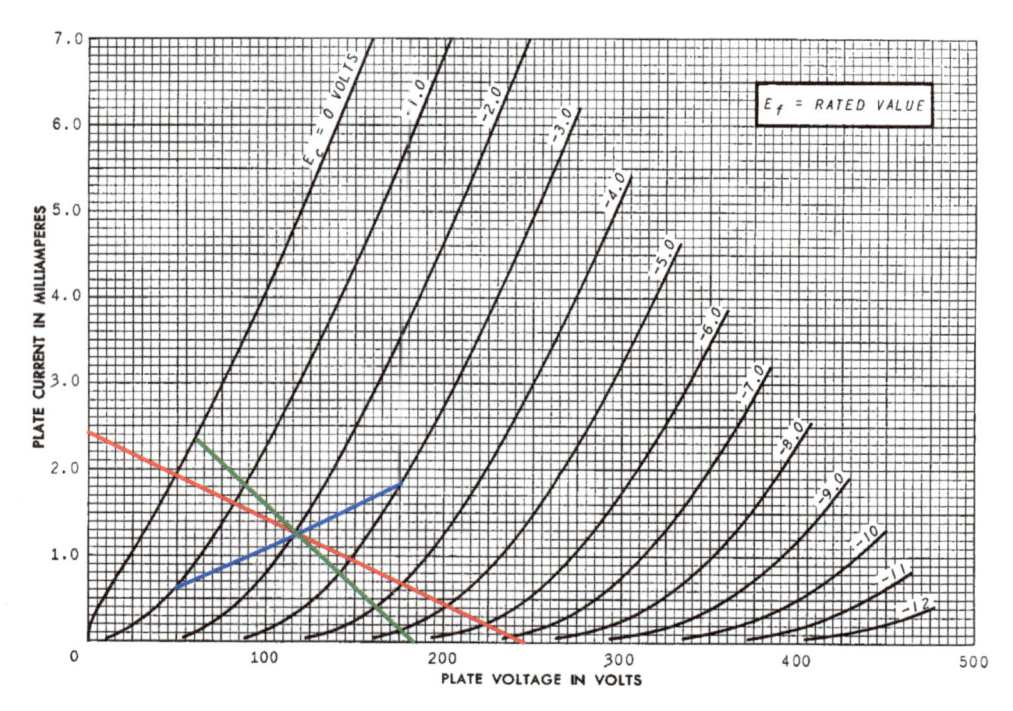

According to Ohm's Law, if the DC grid bias is -1V, -2V, or -3V, the values for plate current passing through the 1.64kΩ cathode resistor are

$$\frac{1V}{1.64k\Omega} = 0.61mA$$

$$\frac{2V}{1.64k\Omega} = 1.22mA$$

$$\frac{3V}{1.64k\Omega} = 1.83mA$$

These values are the endpoints for the two line segments shown in blue. They intersect the load line at the DC operating point: a grid-to-cathode voltage of -2V, a plate-to-cathode voltage of 118V, and a plate current of 1.3mA.

The instrument output A drives volume and tone controls that load down the stage to varying degrees depending on knob position. The greatest AC load (least amount of AC resistance) is at very high frequencies when the capacitors act as short circuits. When the controls are at 50-percent rotation (10-percent resistance) for the instrument channel and the

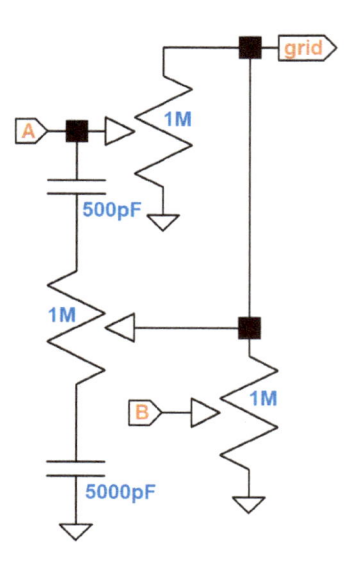

**50 – CHAPTER 2**

microphone channel's volume control is at minimum, the circuit looks like this at high frequencies.

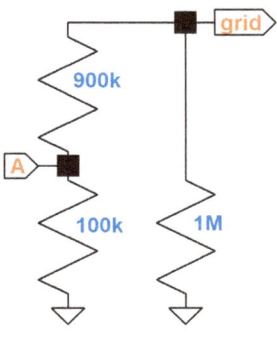

The resistance between node A and ground is 100kΩ in parallel with 1.9MΩ, which is so large by comparison that the parallel resistance is only slightly less than 100kΩ. For audio signals this AC load is in parallel with the 100kΩ plate load resistor for an effective AC plate load of 50kΩ. According to Ohm's Law, when the plate current swings from its DC value of 1.3mA to a value of 0mA, the plate voltage swings by

$$(1.3mA)(50k\Omega) = 65V$$

The plate voltage therefore swings from its DC value of 118V to a value of $118V + 65V = 183V$, so the AC load line includes the DC operating point and the endpoint 183V, 0mA. The resulting AC load line is shown in green. It indicates that the input signal at the grid can swing by up to 2V peak before the onset of severe distortion. Because of the 68kΩ voltage divider, this is only half the signal level measured at the guitar input jack: 4V peak (+9dBV), representing input headroom for the first stage measured at the jack. Without pedal boost, a guitar signal is well below this limit, so the stage is not likely to be overdriven under ordinary conditions.

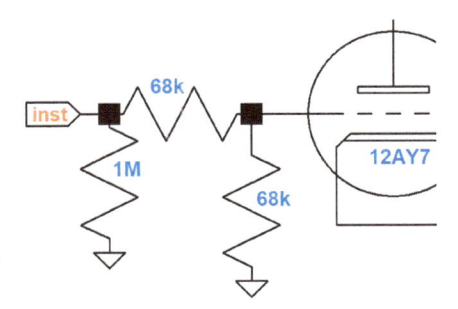

The unloaded voltage gain for a 12AY7 amplification factor of 44, a 12AY7 plate resistance of 25kΩ, a 100kΩ plate load resistor, and a fully bypassed cathode resistor is

$$\frac{(44)(100k\Omega)}{100k\Omega + 25k\Omega} = 35.2 \; (30.9dB)$$

The voltage amplifier's output impedance, which we used in the previous section to determine volume and tone control circuit's insertion loss, is equal to the plate load resistor in parallel with the triode's plate resistance:

$$\frac{1}{\frac{1}{100k\Omega} + \frac{1}{25k\Omega}} = 20k\Omega$$

The unloaded signal level driving the volume and tone control circuit needs to be +5.1dBV for full power when the knobs are at noon. The corresponding signal level at the guitar input jack is reduced by preamp amplification and increased by the 68kΩ resistor network:

$$5.1 dBV - 30.9 dB + 6 dB = -19.8 dBV\ (145 mV\ peak)$$

With the volume and tone controls cranked to maximum the required signal level for full power is

$$-11.8 dBV - 30.9 dB + 6 dB = -36.7 dBV\ (21 mV\ peak)$$

This is well within the capabilities of a single-coil pickup.

The 25µF cathode bypass capacitor is relatively large, so it creates only a tiny amount of bass attenuation. The 0.05µF coupling capacitor forms a high-pass RC filter with the output impedance that drives it and the shunt resistance of the volume control.

At high volume settings, the capacitor's reactance is relatively small compared to the volume control's shunt resistance between the wiper and ground. As the volume control is turned down, however, the shunt resistance decreases and the filter's effective break frequency increases. As a result, as the volume control setting is lowered, bass attenuation increases at a faster rate than the overall volume. This is an inherent feature of a potentiometer whose wiper is used as the signal input instead of the signal output, representing just another contribution to the unique sonic palette of a tweed Deluxe.

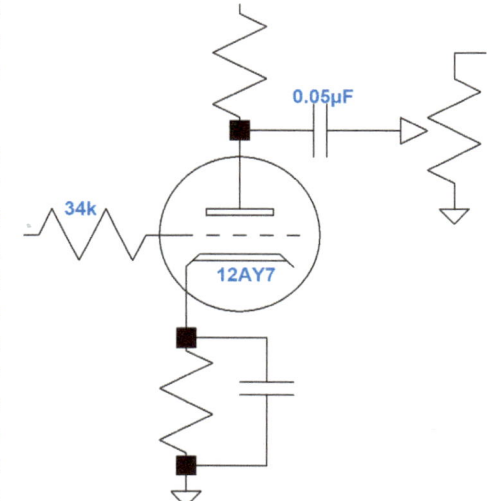

The grid-to-cathode and the grid-to-plate capacitances for a 12AY7 are 1.3pF. We determined that the unloaded voltage gain is 35.2 (30.9dB). Loaded gain is less and depends on knob positions. An upper bound on Miller capacitance is therefore $1.3pF + (35.2 + 1)(1.3pF) = 48pF$.

Ignoring the output impedance of the guitar (i.e., assuming it is a perfect 0Ω), the series resistance in front of the grid is approximately equal to the two 68kΩ resistors in parallel for an effective resistance of 34kΩ.

The -3dB break frequency is

$$\frac{1}{2\pi(34k\Omega)(48pF)} = 98 kHz$$

52 – CHAPTER 2

This is an order of magnitude greater than guitar frequencies so the input circuit causes very little treble attenuation: about 0.1dB at 10kHz.

## System Profile

The schematic on the next page documents what we have determined for the 5D3. The schematic includes

1. DC voltages and DC currents that define the DC operating points,
2. dBV signal levels for a 1kHz signal required to drive the amp to full power with the instrument channel's controls at 50-percent rotation and the microphone volume control at minimum,
3. dBV headroom (HR) at the input to each stage, and
4. dBV ripple from the power supply.

System sensitivity is -19.8dBV (145mV peak). With both controls cranked to maximum, sensitivity is -36.7dBV (21mV peak). These and other signal levels can be visualized by drawing what I call a *system profile*.

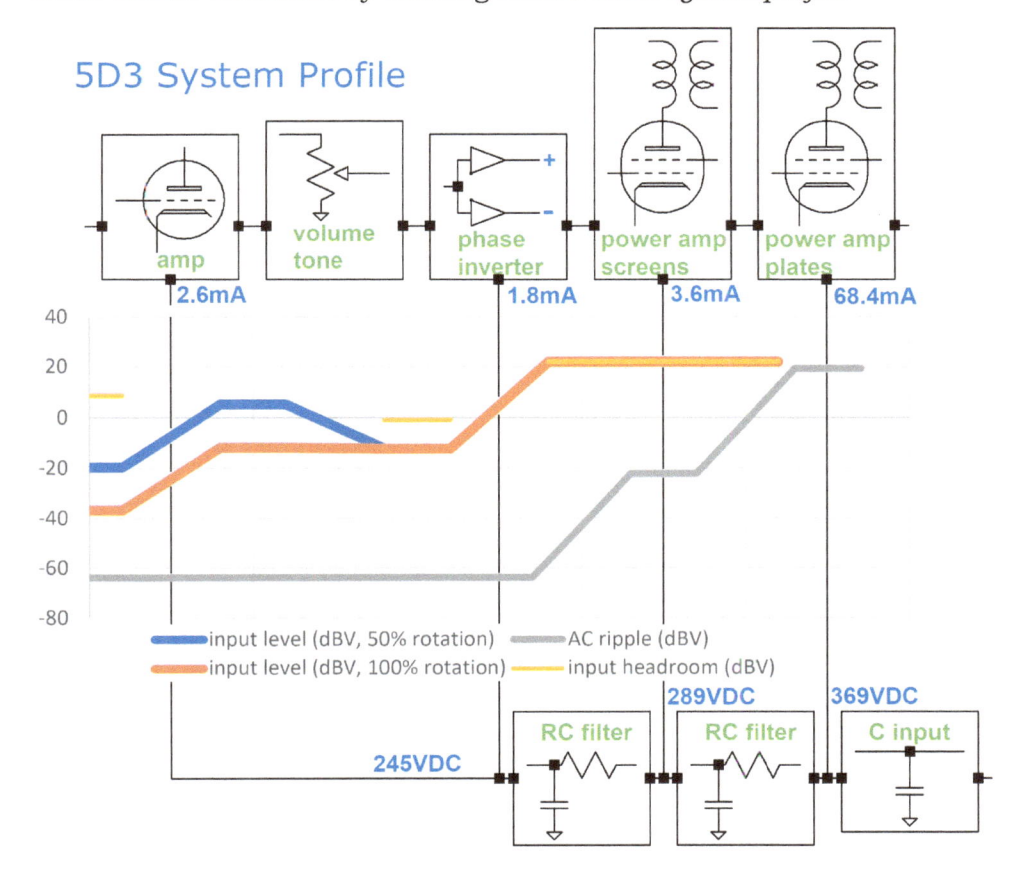

### 5D3 System Profile

| Stage | amp | volume tone | phase inverter | power amp screens | power amp plates |
|---|---|---|---|---|---|
| Current | 2.6mA | | 1.8mA | 3.6mA | 68.4mA |

Legend:
- input level (dBV, 50% rotation)
- input level (dBV, 100% rotation)
- AC ripple (dBV)
- input headroom (dBV)

Power supply: 245VDC → RC filter → 289VDC → RC filter → 369VDC → C input

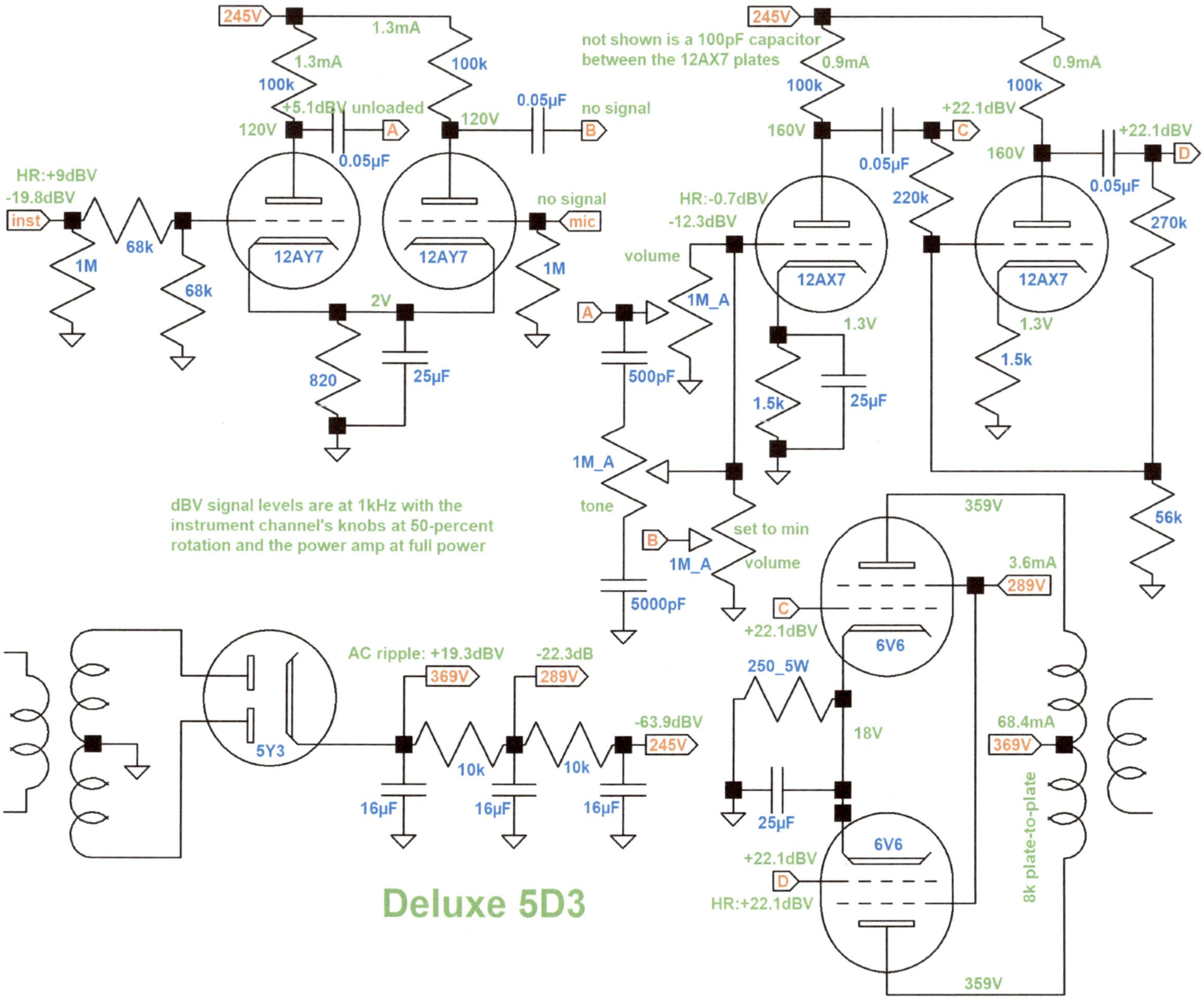
Deluxe 5D3
dBV signal levels are at 1kHz with the instrument channel's knobs at 50-percent rotation and the power amp at full power
not shown is a 100pF capacitor between the 12AX7 plates
245V
1.3mA
1.3mA
100k
100k
+5.1dBV unloaded
120V
120V
0.05µF
0.05µF
A
B
no signal
HR:+9dBV
-19.8dBV
inst
68k
68k
1M
12AY7
12AY7
no signal
mic
1M
2V
820
25µF
245V
0.9mA
0.9mA
100k
100k
160V
160V
+22.1dBV
+22.1dBV
C
D
0.05µF
0.05µF
220k
270k
HR:-0.7dBV
-12.3dBV
volume
1M_A
12AX7
12AX7
1.3V
1.3V
1.5k
1.5k
25µF
500pF
1M_A
tone
5000pF
set to min
volume
1M_A
56k
359V
3.6mA
289V
+22.1dBV
6V6
6V6
250_5W
18V
25µF
+22.1dBV
HR:+22.1dBV
359V
68.4mA
369V
8k plate-to-plate
5Y3
AC ripple: +19.3dBV
369V
-22.3dB
289V
-63.9dBV
245V
10k
10k
16µF
16µF
16µF

The blue trace shows the signal level at the input to each stage at full power with the volume and tone controls at 50-percent rotation. The orange trace is with the controls at maximum. The yellow traces show input headroom for the preamp, phase inverter, and power amp. The gray trace shows AC ripple for the plate and screen supplies. All signal levels are in dBV.

The profile shows typical patterns for a guitar amplifier at full power:

1. The signal level at the guitar input jack matches input sensitivity.
2. The signal level at the power amp input matches its input headroom.
3. Signal levels generally zigzag their way higher, getting closer to headroom limits and driving the tubes harder with each stage. The first-stage preamp runs squeaky clean while the power amp is on the cusp of overdrive.

For the power amp plate supply, AC ripple is nearly as great as the input signal level. This is mitigated by the fact that the plates are less susceptible to ripple than the screens. Moreover, the Deluxe has a push-pull design, so ripple applied equally to both 6V6 plates cancels itself.

## System Voicing

We will define *system voicing* as the relative frequency response from the input jack to the power amp grids. A SPICE AC analysis simulation shows that it is relatively flat compared to amps with a Bassman-style stack, which has considerable middle scoop. With the knobs at noon here is the frequency response from input jack to the power amp grids relative to the response at 820Hz.

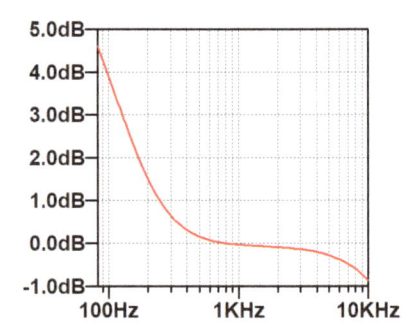

Below 82Hz the response eventually falls off severely, a desirable effect that is caused mostly by coupling capacitors.

## Evolutionary Development Continues

As a successful system design, the 5D3 serves as the starting point for modifications incorporated into the highly regarded 5E3. Big changes are in the works as Fender transitions its product line from wide panels to narrow panels. In the next chapter we will see what rolled out of Fullerton, California for the next iteration of the Deluxe.

# Chapter 3: Pinnacle – The Narrow Panel Deluxe 5E3

This could easily be the final chapter of my book. After all, what more is there to say after examining one of the most iconic guitar amps of all time?!

> "Ah, the glorious narrow-panel 5E3 tweed Fender Deluxe. This is such a ubiquitous design that there's no point detailing it. Hell, it's in our blood, part of our shared consciousness, hardwired into every guitarist's genetic memory."[67] –Dave Hunter

The 5E3 represents a pinnacle of guitar amplifier system design that starts with the woody and progresses through the TV-front and wide panels. The basic architecture for the signal chain and power supply is established by the Model 26. The volume and tone controls are integrated in the 5A3. The conversion to miniature triodes is carried out by the 5D3, which serves as a baseline for the circuit of this chapter.

## System Design Concept

> "The Tweed Deluxe is such a seminal amplifier, is so desirable and (in its original form) so expensive, that there are at least 30 or 40 companies out there making clones or variants of it, either as kits or as completed amplifiers. ... Part of the reason for this plethora of plagiarism is that the 5E3 circuit is extremely simple, and so anyone who can hold a soldering iron can probably build one fairly easily."[68] –Bob Thomas

Fender's schematic and layout for the 5E3 do not include DC voltage measurements, so the DC voltages and currents on the schematic on the next page are my estimates using 5Y3 and 6V6 data sheets, SPICE simulation, and Ohm's Law.

The new 5E3 is caught up in Fender's mid-1950s burst of circuit refinements designed to coax more power from its existing models. The high-end Twin 5E8 evolves the most in this direction:[69]

1. Parallel 5U4 rectifiers replace the Twin 5D8's 5Y3 rectifiers. This reduces sag, raising the power amp's supply voltages.

---

[67] Dave Hunter, **Amped**, (Minneapolis: Voyageur Press, 2012), p. 64.
[68] Bob Thomas, "Torres Engineering", **Sound on Sound**, February 1, 2014. Available at https://web.archive.org/web/20140201102001/http://www.soundonsound.com/pm/nov07/articles/torrestweed.htm (Retrieved November 24, 2020)
[69] John Teagle and John Sprung, **Fender Amps: The First Fifty Years**, (Milwaukee: Hal Leonard, 1995), p. 81.

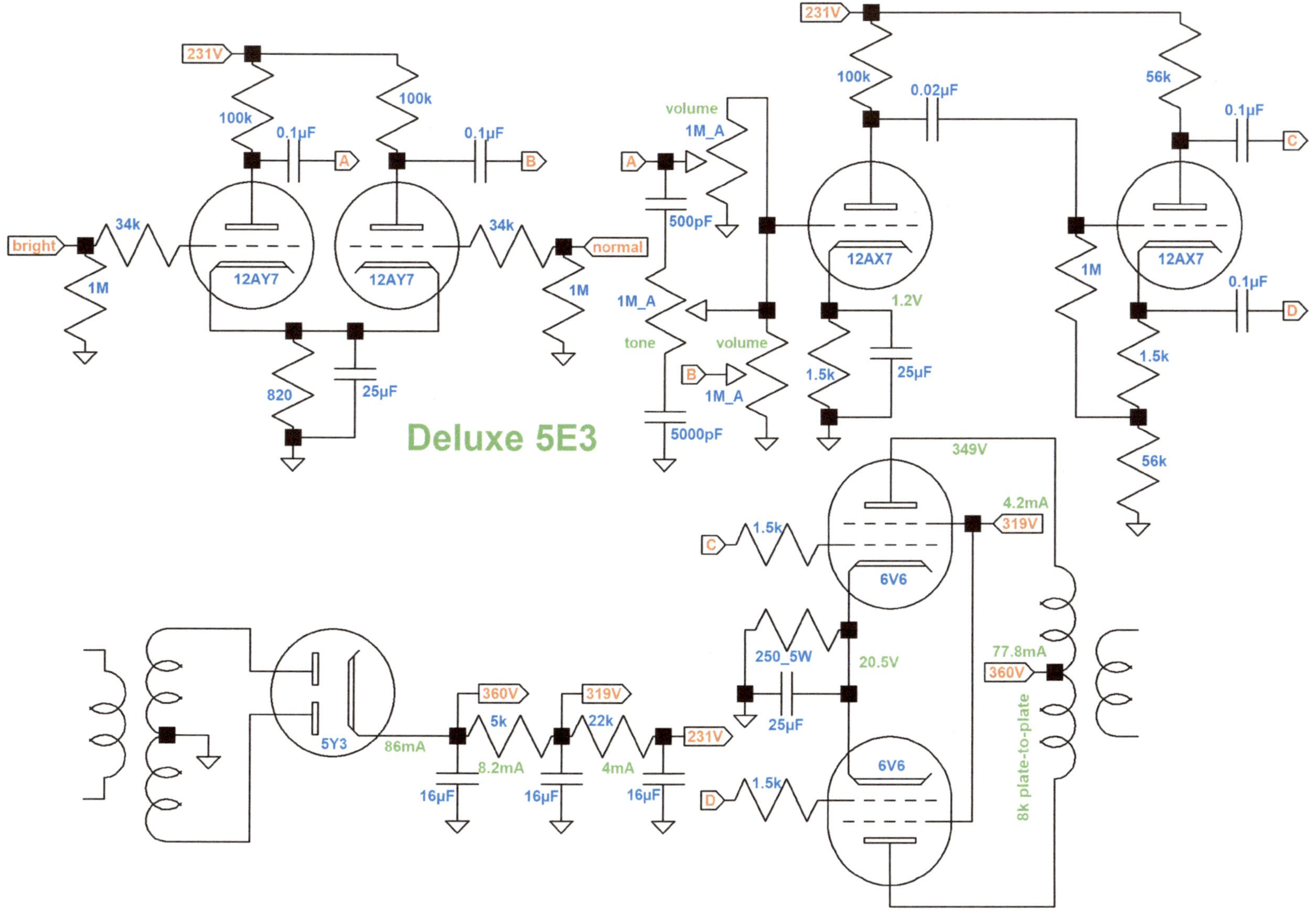
Deluxe 5E3
bright
34k
1M
231V
100k
0.1µF
A
12AY7
100k
0.1µF
B
12AY7
34k
normal
1M
820
25µF
volume
1M_A
A
500pF
1M_A
tone
5000pF
volume
B
1M_A
231V
100k
0.02µF
12AX7
1.2V
1.5k
25µF
56k
0.1µF
C
12AX7
1M
0.1µF
D
1.5k
56k
349V
4.2mA
319V
C
1.5k
6V6
250_5W
20.5V
25µF
6V6
D
1.5k
77.8mA
360V
8k plate-to-plate
5Y3
86mA
360V
5k
8.2mA
319V
22k
4mA
231V
16µF
16µF
16µF

2. A new power transformer is incorporated to create a fixed bias supply. This eliminates the voltage drop across the cathode resistor, making the 6L6 screen-to-cathode voltage equal to the full screen supply voltage.
3. An expensive choke replaces the 5D8's 10kΩ RC ripple filter resistor, raising the effective power amp supply voltages further.
4. The 5D8's feedback resistor value is doubled for less negative feedback, creating more voltage gain.
5. Finally, the 12AX7 triodes for the 5D8 self-balancing phase inverter are refashioned into a voltage amplifier and a split-load phase inverter, eliminating the intervening attenuator.

The new Twin design represents a complete overhaul using high-end upgrades. Changes in store for the Deluxe 5E3, on the other hand, are less dramatic and considerably less expensive. The tweed Deluxe continues to use a 5Y3 rectifier. Its cathode bias for the power amp also remains intact – fixed bias for the Deluxe comes years later. Like the Twin, however, the 5E3 gets a split-load phase inverter and a separate gain stage. Instead of an expensive choke, ripple filter resistor values are tweaked to get more screen voltage and the same level of ripple at the front end.

The Deluxe 5E3 is one of the first amps to get an improved input circuit: a gamma network to replace the 68kΩ voltage divider. The guitar signal is now amplified before being subject to attenuation, a prudent design practice that carries forward to today's amps.

Like the Twin, the aggressive attenuator between the phases of the phase inverter is eliminated by separating voltage amplification from phase inversion. The Deluxe 5D3 has this signal path from the jack to the power amp grids:

attenuator ➡ amp ➡ controls ➡ amp ➡ attenuator ➡ amp

In contrast, the narrow-panel 5E3 signal chain is as clean as it gets for push-pull:

amp ➡ controls ➡ amp ➡ phase inverter

Before we crunch some numbers, here is an overview of circuit enhancements to be examined in the sections ahead.

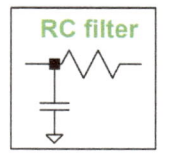

**RC filter** — 6V6 screen voltages are increased by adjusting ripple filter resistor values.

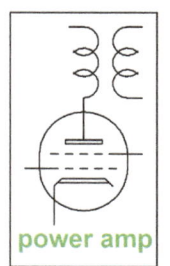

**power amp** — Higher 6V6 screen voltages translate to greater output power.

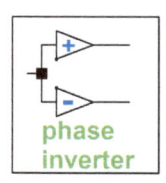

**phase inverter** — A split-load phase inverter is introduced.

**preamp** — The first triode of the 5D3 paraphase becomes a second-stage voltage amplifier.

## Coaxing More Power from the Power Amp

In the mid-1950s Fender is focused on greater power before breakup, i.e. more clean headroom.

> "When it comes to warm, harmonic distortion produced at moderate volumes in a cranked tube amp, Leo Fender may have already built the ultimate tone machines by, say, '53 or '54. But he never intended for his amps to be played consistently at full volume, and despite his idiosyncrasies and departures from convention, in one respect he was like other amp designers of this time: for him, distortion was the enemy, headroom the goal."[70] –Tom Wheeler

On the next page in red is the relationship between net plate current and plate-to-cathode voltage that we observed in the last chapter for the 5D3. The load line intersects an imagined curve for 271V screen-to-cathode (red

---

[70] Tom Wheeler, **The Soul of Tone**, (Milwaukee: Hal Leonard, 2007), p. 181.

dot) at a plate voltage 105V and a plate current of 116mA for an output power of approximately 13.7W, assuming the supply voltages do not sag.

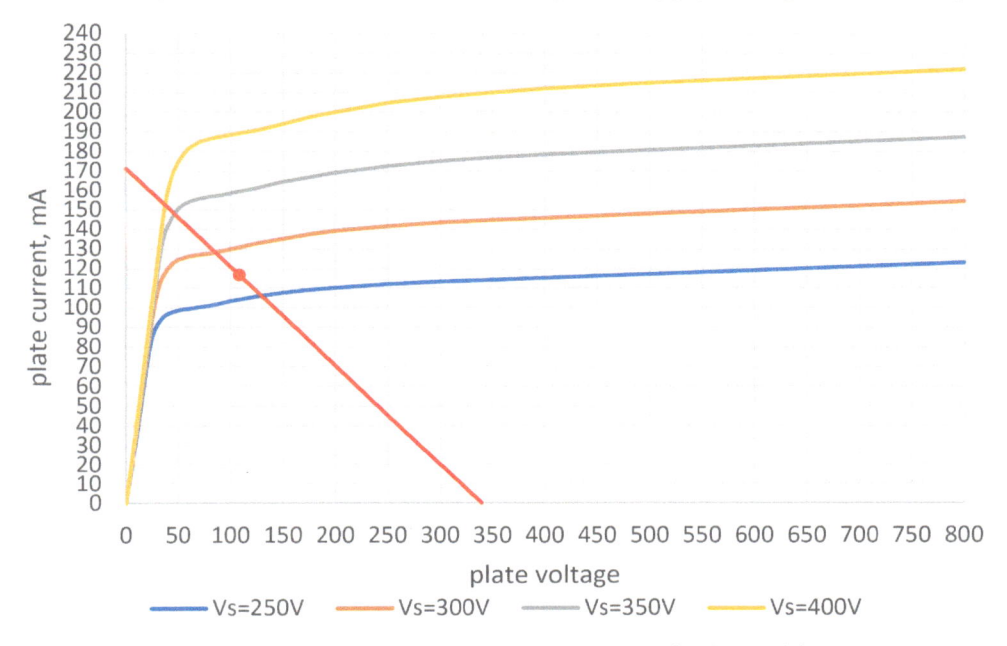

If the screen supply voltage increases, the amp's 0V grid curve moves upward. A higher screen voltage creates more plate current, which causes power supply sag, reducing the plate supply voltage, so the load line moves to the left, intersecting the curve closer to the knee. The net result is greater plate current and plate voltage swing for a net increase in power.

Fender increases the screen supply voltage by reducing the RC ripple filter resistor value between the plate and screen supplies.

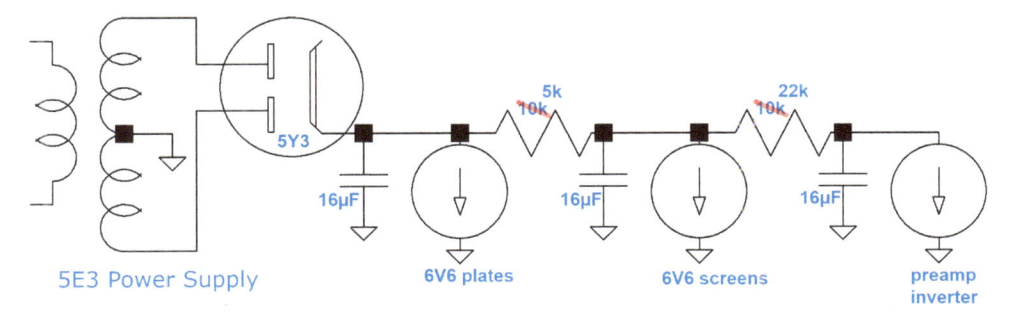

This reduces ripple filtering for the screen supply. To maintain the same level of ripple filtering for the triode supply, the DC resistance between the screen and triode supplies is increased. The net result is more hum at the screens and approximately the same hum everywhere else.

According to Ohm's Law, reducing the first RC resistor value by half decreases the voltage drop across it by half. If nothing else changes, the 5D3's 80V drop becomes 40V, so the 289V screen supply voltage increases to 329V. The change has repercussions, however – a higher screen voltage creates more 6V6 plate current which creates more power supply voltage sag. We therefore get something less than 329V for the screen supply.

## Power Amp Performance

The 5E3 adds 1.5kΩ grid-stopper resistors to the power tubes to prevent radiofrequency interference and parasitic oscillation.

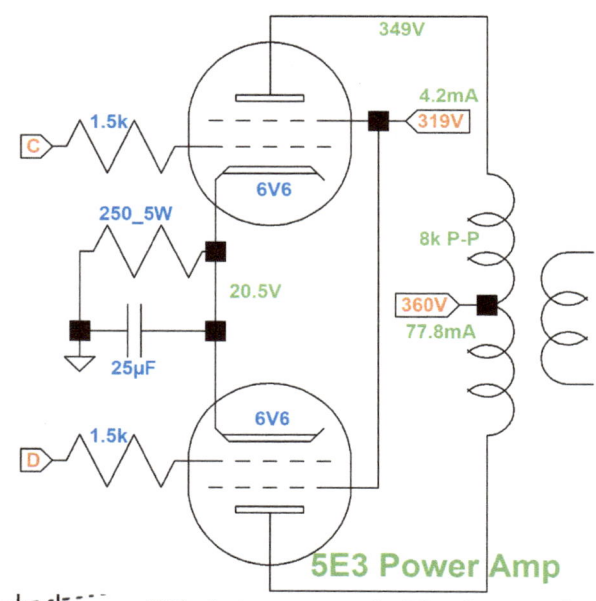

Fender's layout shows that the resistors are soldered directly to the tube sockets, as shown below. This eliminates the inductance introduced by any wire between the resistor and the grid. For radiofrequency suppression the resistor and tube electrodes form a simple RC low-pass filter without any extraneous parasitic impedances.

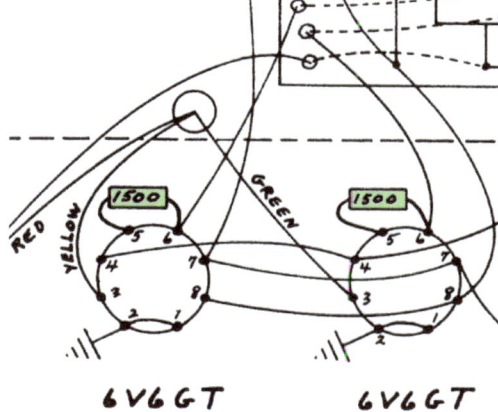

### 5E3 Power Amp

DC plate current is 39mA per tube at 328V plate-to-cathode for a plate dissipation equal to $(328V)(39mA) = 12.8W$, slightly more than the data sheet's limit of 12W. Screen current per tube is 2.1mA at 299V screen-to-cathode for a screen dissipation of $(299V)(2.1mA) = 628mW$, which is well within the data sheet maximum of 2W. Screen dissipation is more of a concern at full power, however. For the 5E3, screen current averages about 7mA at full power, as we will see later in this chapter.

If the plate-to-cathode voltage swings from 328V to 0V across an effective 2kΩ transformer primary impedance (8kΩ plate-to-plate), net plate current swings to

$$\frac{328V}{2k\Omega} = 164mA$$

The endpoints for the AC load line are therefore 0V, 164mA and 328V, 0mA, as shown here by the red line.

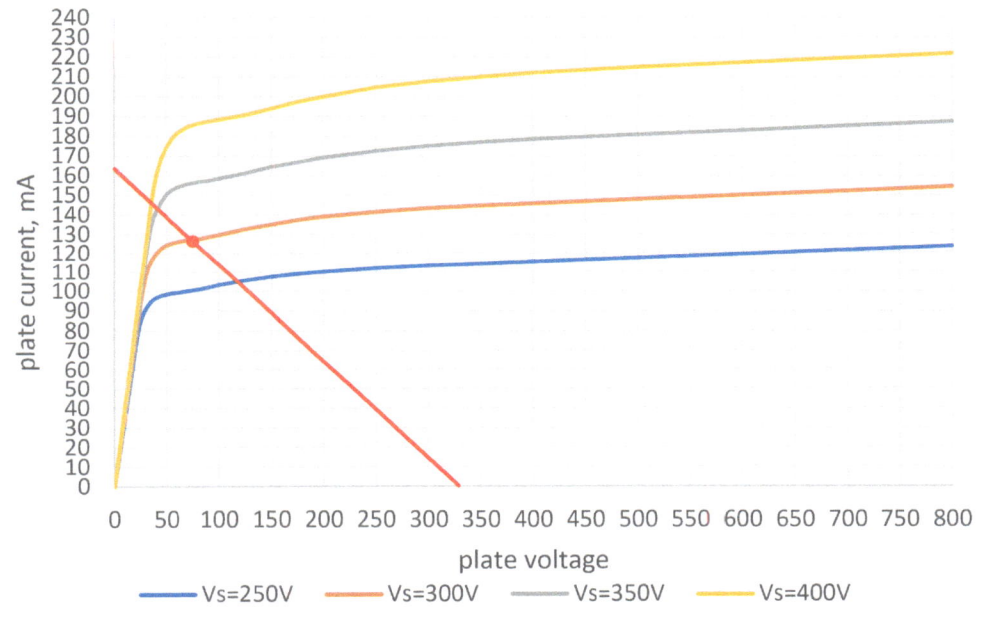

The load line intersects an imagined 0V curve for a 299V screen at 75V, 128mA (depicted by the red dot) for an output power of approximately

$$\left(\frac{1}{2}\right)(328V - 75V)(128mA) = 16.2W$$

This assumes the power supply voltages do not sag at full power. Our 13.7W estimate for the 5D3, on the other hand, also makes this assumption, so comparing the two values makes sense.

Based on these results, the 5E3 has 18-percent more power. A glossy product brochure notwithstanding, this can appear to be more substantial than it really is. A doubling of power is an increase of 3dB, not 6dB as it is with a ratio of voltages. In terms of decibels, the Deluxe power increase is only

$$10 \log\left(\frac{16.2W}{13.7W}\right) = 0.7dB$$

**PINNACLE – THE NARROW PANEL DELUXE 5E3**

This is on the edge of perceptibility. Nevertheless, in the 1950s increased power is the direction Fender is taking with push-pull designs. Moreover, the 5E3 power increase is achieved at no cost – the modifications do not require a new transformer, a stiffer rectifier, fixed bias, or an expensive choke.

Input headroom matches the DC cathode voltage: 20.5V peak (+23.2dBV). This is the signal level required at the power amp grids to achieve full power.

## Power Supply Ripple

To compensate for less ripple filtering in the screen supply, the second RC ripple filter resistor value is slightly more than doubled to 22kΩ. This approximately doubles the difference between the screen and triode supply voltages.

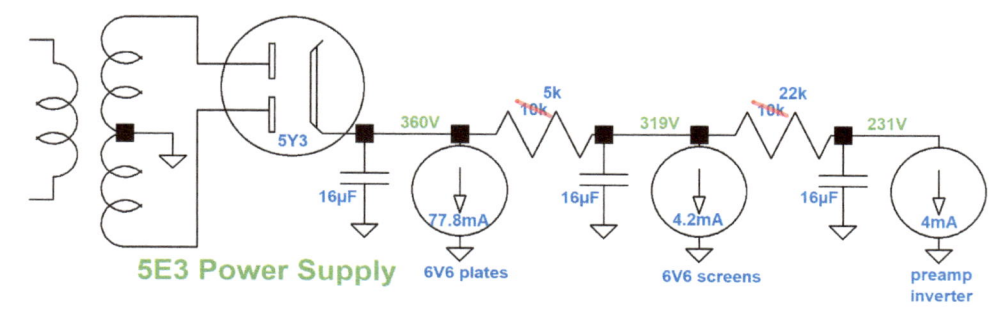

The total DC load on the power supply is 86mA. For a DC load of 86mA at 360V, a 16μF capacitor input, and a 120Hz ripple frequency, the ratio of the ripple voltage in RMS to DC voltage is[71]

$$\frac{\sqrt{2}(86mA)}{2\pi(120Hz)(360V)(16\mu F)} = 0.028$$

This means the ripple level is $(0.028)(360V) = 10V\ RMS$, which is 14V peak (+20dBV).

Ripple attenuation for the first RC filter, with a 5kΩ resistor and 16μF capacitor, is

$$\frac{1}{\sqrt{[2\pi(120Hz)(5k\Omega)(16\mu F)]^2 + 1}} = 0.0166\ (-35.6dB)$$

AC ripple at the 6V6 screens is therefore

---

[71] **Silicon Rectifier Handbook**, (Bloomington: Sarkes Tarzian, 1960), pp. 41, 42.

$$20dBV - 35.6dB = -15.6dBV \ (235mV \ peak)$$

Ripple attenuation for the second RC filter, with a 22kΩ resistor and 16µF capacitor, is

$$\frac{1}{\sqrt{[2\pi(120Hz)(22k\Omega)(16\mu F)]^2 + 1}} = 0.00377 \ (-48.5dB)$$

AC ripple for the triode supply is

$$-15.6dBV - 48.5dBV = -64.1dBV \ (882\mu V \ peak)$$

## A New Voltage Amplifier and Split-Load Phase Inverter

The first stage of a paraphase is a simple voltage amplifier. It becomes part of a "phase inverter" when it is followed by an attenuator and another voltage amplifier. The attenuator-amplifier combination creates an inverting voltage "amplifier" with unity gain.[72] The 5E3 eliminates the attenuator and second amplifier and substitutes a split-load phase inverter.

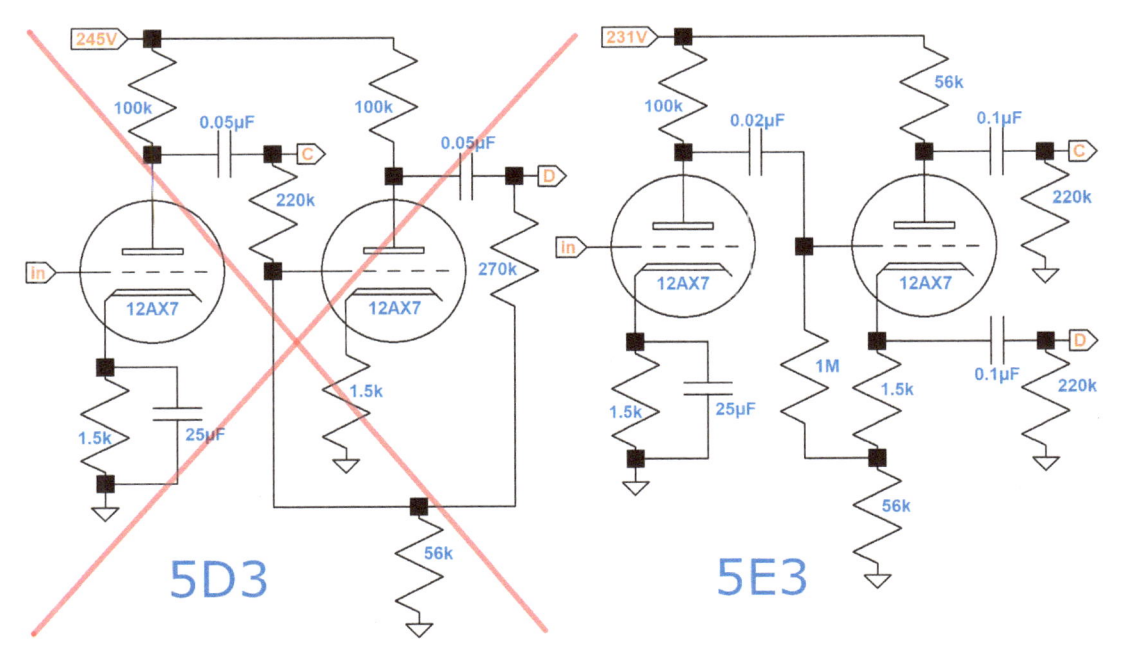

It also has unity gain but provides both phases to drive the power amp. The first triode circuit remains a voltage amplifier, just like before, but it is no longer part of the phase inverter.

---

[72] Richard Kuehnel, **Guitar Amplifier Electronics: Basic Theory**, (Seattle: Amp Books, 2018), pp. 126-130.

The Bassman 5D6, Super 5E4-A, Pro 5E5, Bandmaster 5E7, and the Twin 5E8 get the same treatment.

> "The [5E3] circuit runs at higher voltages than other models and features a split-phase inverter and driver that add a little gritty breakup at the start of the output stage."[73]

## Voltage Amplifier Performance

For the voltage amplifier, the DC load line shown in red joins the DC plate-to-cathode voltage of 231V on the X axis to a Y-axis plate current of

$$\frac{231V}{100k\Omega + 1.5k\Omega} = 2.28mA$$

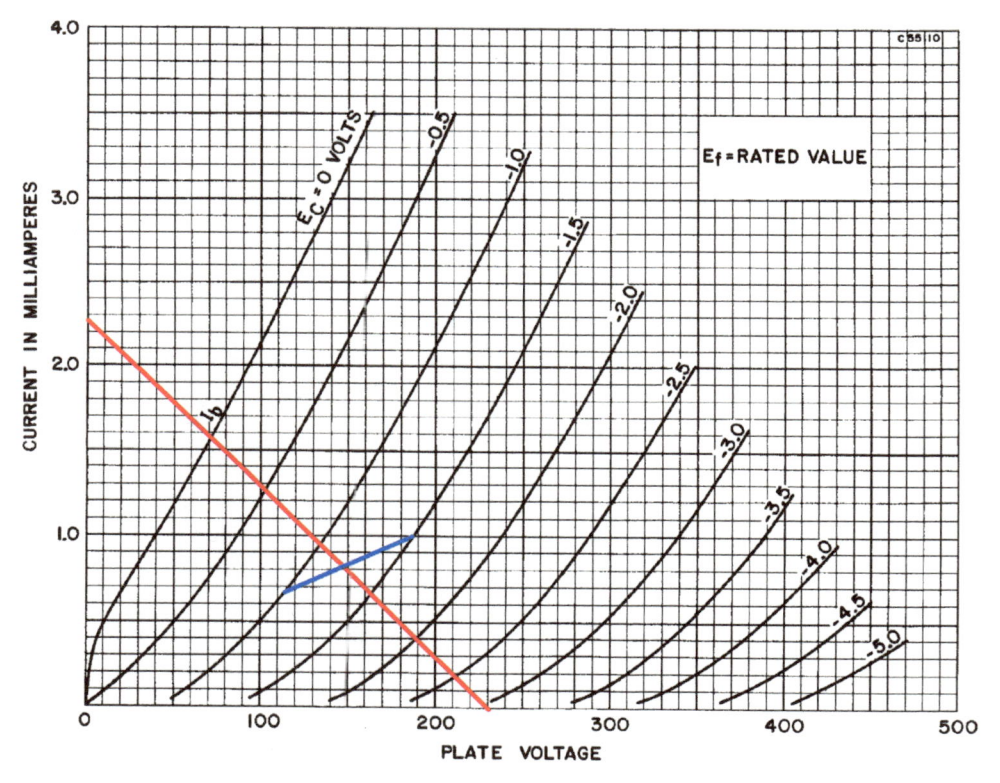

If the grid voltage is -1V or -1.5V, then the plate current is

$$\frac{1V}{1.5k\Omega} = 0.67mA$$

---

[73] Damian Fanelli, Christopher Scapelliti and Tom Gilbert, "The 10 Most Iconic Guitar Amps of All Time," **Guitar World**, April 24, 2020.

**66 – CHAPTER 3**

$$\frac{1.5V}{1.5k\Omega} = 1mA$$

These values define the blue line segment. The DC operating point is defined by the intersection: a grid-to-cathode voltage of -1.2V, a plate-to-cathode voltage of 149V, and a plate current of 0.8mA.

The split-load phase inverter represents a very light AC load, so the AC load line is almost identical to the DC load line. The grid can swing by 1.2V before breaking into distortion, so input headroom for the voltage amplifier is 1.2V peak (-1.4dBV).

The unloaded voltage gain for a 12AX7 amplification factor of 100, a plate resistance of 62.5kΩ, a 100kΩ plate load resistor, and a fully bypassed cathode resistor is

$$\frac{(100)(100k\Omega)}{100k\Omega + 62.5k\Omega} = 61.5 \; (35.8dB)$$

Because the voltage amplifier drives a light AC load, its loaded voltage gain is almost the same.

## Phase Inverter Performance

The phase inverter's DC operating point is nearly identical.

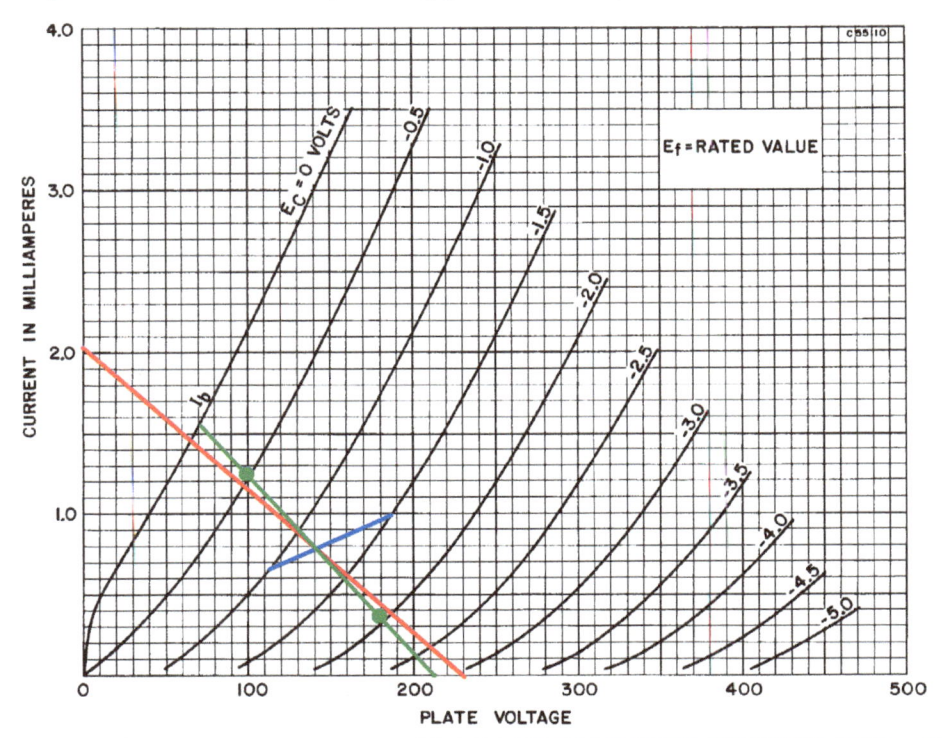

**PINNACLE – THE NARROW PANEL DELUXE 5E3**

The DC load line, shown in red, joins the DC plate-to-cathode voltage of 231V on the X axis to a Y-axis plate current of

$$\frac{231V}{56k\Omega + 1.5k\Omega + 56k\Omega} = 2.04mA$$

The blue grid line is the same as for the voltage amplifier. The DC operating point is defined by the intersection: a grid-to-cathode voltage of -1.2V, a plate-to-cathode voltage of 140V, and a plate current of 0.8mA.

For an AC signal, each 56kΩ resistor is in parallel with a 220kΩ grid-leak resistor for an AC load of 45kΩ. (The extra 1.5kΩ at the cathode, which sets the DC grid bias, is small enough to be ignored.)

If the plate current swings from the DC operating point value of 0.8mA to a value of 0mA, then the plate-to-ground voltage swings positive by

$$(0.8mA)(45k\Omega) = 36V$$

The cathode-to-ground voltage swings negative by the same amount, so the plate-to-cathode voltage swings positive by 2(36V) = 72V. The green AC load line therefore contains the endpoint defined by 0mA and 140V + 72V = 212V. The slope of the AC load line is

$$\frac{-72V}{0.8mA} = -90k\Omega$$

This makes sense, because for audio signals the plate and cathode loads are each 45kΩ. The triode is therefore in series with $45k\Omega + 45k\Omega = 90k\Omega$.

According to the AC load line, plate current can swing to a maximum of 1.55mA. The tube's heater is at ground potential, so cathode-to-heater voltage can swing to

$$(1.55mA)(56k\Omega + 1.5k\Omega) = 89V$$

This is less than the 180V limit for a 12AX7. Nevertheless, elevated cathode voltages for this type of phase inverter can create fear and doubt for mid-20th-century circuit designers.[74] For the next design iteration, the Deluxe shifts to a long-tailed pair, which lowers and steadies the cathode voltage, reduces the risk of heater-to-cathode leakage,[75] and maybe allows certain residents of Fullerton California to sleep easier at night.

At full power, the plate-to-ground voltage swings by 20.5V peak, matching the input headroom of the power amp.

---

[74] Personal correspondence with Paul Reid, February 2021.
[75] F. Langford-Smith, **Radiotron Designer's Handbook**, 4th ed., (Harrison: RCA, 1952), p. 81.

The cathode-to-ground voltage swings by 20.5V peak with opposite phase, so the plate-to-cathode voltage swings by 41V peak. This extent of swing is marked by the green dots on the AC load line. According to the AC load line, the grid-to-cathode voltage swings by about 0.7V peak.

Maximum plate-to-cathode voltage swing is about 72V peak, representing 36V peak at the plate output and the cathode output, so at full power the phase inverter operates slightly more than halfway to overdrive.

Larger amps in Fender's lineup (Bassman 5D6, Super 5E4-A, Pro 5E5, Bandmaster 5E7, and the Twin 5E8) get the same split-load phase inverter. A maximum swing of 36V peak at each output seems mighty small for driving their big bottles. The key difference is that the phase inverters in these amps operate from much higher plate supply voltages, which increases the limits of voltage swing. For the 5E3, a 231V plate supply is quite sufficient – the power amp transitions into overdrive long before the phase inverter reaches its limits.

Since the phase inverter has approximately unity gain, the maximum grid-to-ground voltage swing is approximately the same as the maximum plate-to-ground voltage swing: 36V peak. Input headroom is therefore 36V peak (+28.1dBV). Gain for the voltage amplifier and phase inverter together is about 35.4dB, only slightly less than the 35.8dB that we calculated for the voltage amplifier unloaded gain. There is only about 0.2dB imbalance between the plate output (red) and the cathode output (blue).

The 6V6 grids need +23.2dBV to reach full power. The signal level needed at the voltage amplifier grid is

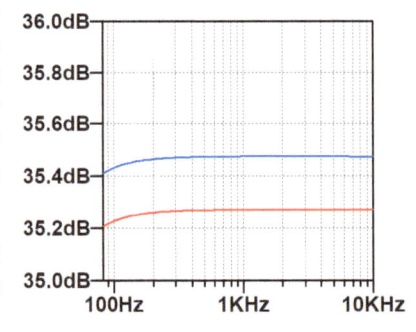

$$23.2 dBV - 35.4 dB = -12.2 dBV \ (347 mV \ peak)$$

## Volume and Tone Control Interactivity

The 5E3 abandons the concept of "instrument" and "microphone" inputs. It now has a "bright" channel and a "normal" channel. Here is the normal channel's frequency response from the high-gain jack to the second-stage grid with all controls at 50-percent rotation (10-percent resistance). On the next page is the bright channel under identical conditions.

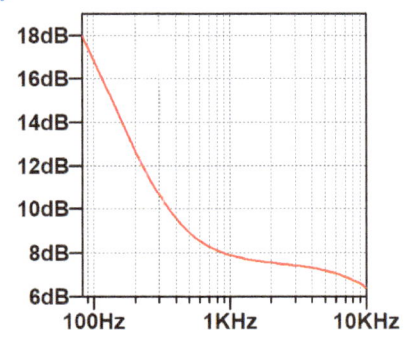

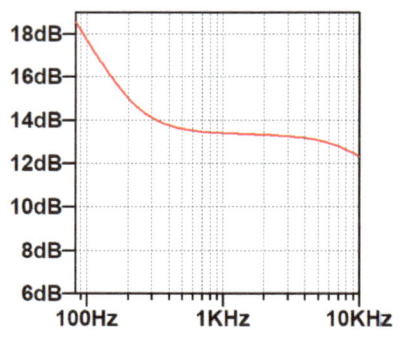

Both channels have substantial bass boost, which happens to include the fundamental of AC ripple for a full-wave rectifier: 120Hz.

"This doesn't help the perceived level of hum one little bit."[76] –Bob Thomas

In effect, the volume controls mix the guitar signal and a 0V audio signal from the unused channel, so when the unused channel's volume control is turned up, active channel gain decreases. Here, for example, is the frequency response from the guitar input jack to the second-stage grid for the bright channel with the normal channel volume control at minimum (red trace) and maximum (blue trace). The other controls are at noon.

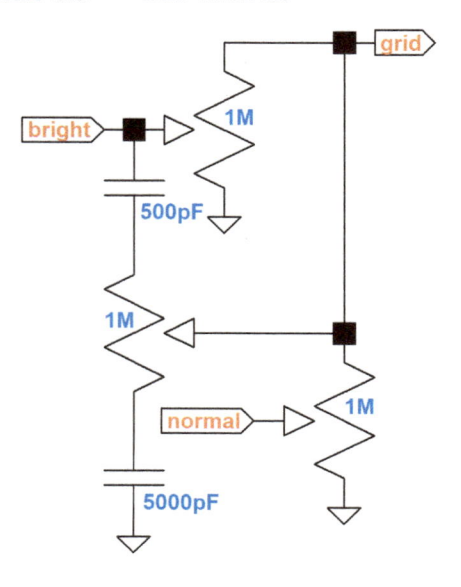

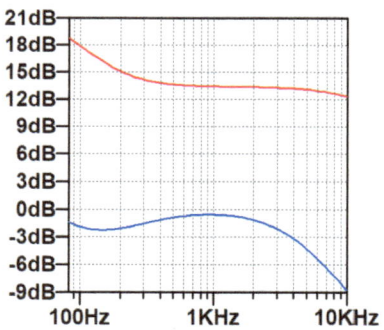

The "tone" control primarily affects treble, but its influence extends all the way down to 82Hz. It therefore affects both volume and tone.

Here, for example, is the bright-channel response with the its volume at noon, the normal channel volume at minimum, and the tone control at minimum (red trace) and maximum (blue trace).

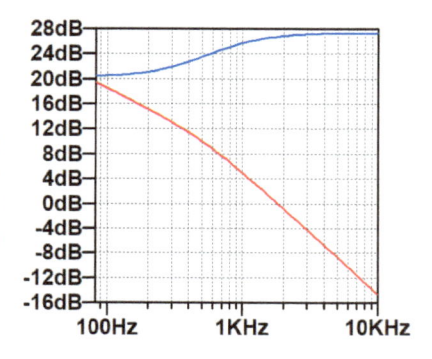

---

[76] Bob Thomas, "Torres Engineering," **Sound on Sound**, February 1, 2014. Available at https://web.archive.org/web/20140201102001/http://www.soundonsound.com/pm/nov07/articles/torrestweed.htm (Retrieved November 24, 2020)

"When you wind the Tone control up towards 12, as well as adding potentially ear-shredding treble, you'll start to add gain. This becomes especially obvious at the higher settings."[77]

In summary, the three controls are highly interactive.

"This is not a bad thing. The various interactions can be complicated, but the guitarist willing to explore them will learn to play his or her amplifier with more subtlety, just as an organist might use the drawbars on a Hammond B-3. As a matter of fact, the interactivity of controls was part of Leo Fender's plan from the very beginning. Remember, he expected his amps to be used not only by instrumentalists but by vocalists, too, often together."[78] –Tom Wheeler

"You can use the volume control of the input that you're not plugged into, together with the Tone control, to modify tone and gain in the other channel. For example, if you're in the Bright channel, you can use the Normal channel Volume pot to affect the Bright channel's mid and gain. Turning up the Tone adds overall gain as well as treble, since it directly drives the second preamp stage. So, balancing the two Volume controls, the Tone control, the volume and tone settings on your guitar — plus your pick and/or finger attack — gives you access to a multitude of distorted sounds. The Tweed Deluxe isn't an amplifier that you plug into and instantly achieve the sound of your dreams. You need to work with it pretty intensively to find out how your particular amp behaves and what it can actually do."[79] –Bob Thomas

With the normal channel's volume at minimum and the other controls at noon, insertion loss at 1kHz for the 5E3 bright channel is the same as for the 5D3: 17.4dB. The unloaded signal level at the driving stage output needed for full power is equal to the level at the second-stage input plus insertion loss:

$$-12.2 dBV + 17.4 dB = +5.2 dBV\ (2.6V\ peak)$$

With the bright channel voltage and tone controls at maximum, insertion loss is about 0.5dB and the required signal level is 16.9dB less:

$$-12.2 dBV + 0.5 dB = -11.7 dBV\ (368mV\ peak)$$

---

[77] Bob Thomas, "Torres Engineering," **Sound on Sound**, February 1, 2014. Available at https://web.archive.org/web/20140201102001/http://www.soundonsound.com/pm/nov07/articles/torrestweed.htm (Retrieved November 24, 2020)
[78] Tom Wheeler, **The Soul of Tone**, (Milwaukee: Hal Leonard, 2007), p. 179.
[79] Bob Thomas, "Torres Engineering".

## Preamp Adjustments

The 5E3 doubles the coupling capacitor values to 0.1µF, an inconsequential change that makes the already flat bass response even flatter. The 5D3's 0.05µF coupling capacitors create only about 0.1dB attenuation at 82Hz. For the 5E3 bass attenuation is even less.

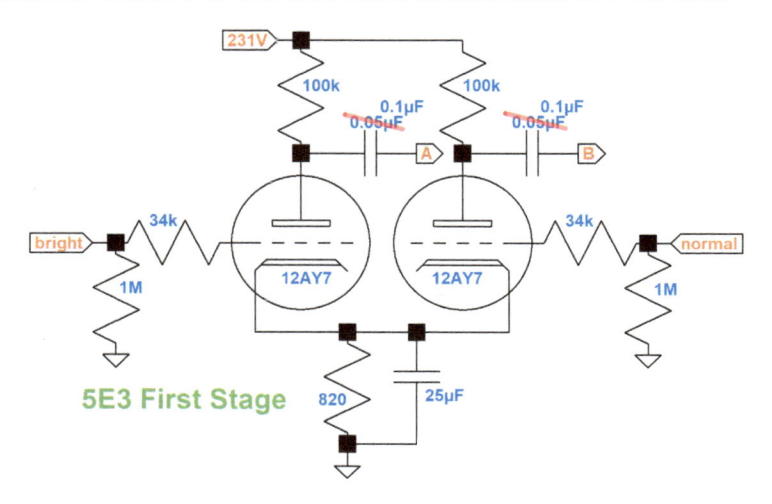

Compared to the 5D3, DC operating conditions shift slightly because of the lower plate supply voltage.

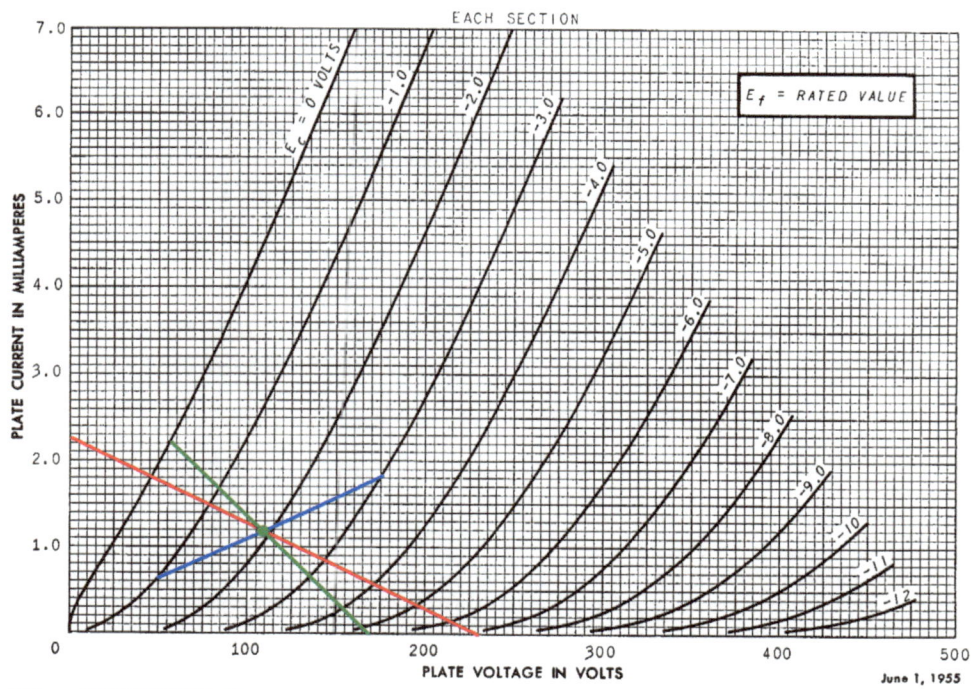

72 – CHAPTER 3

The intersection of the red DC load line and the blue grid line define the DC operating point: a grid-to-cathode voltage of -1.9V, a plate-to-cathode voltage of 108V, and a plate current of 1.2mA.

Like for the 5D3, the worst-case effective AC plate load resistance with knobs at noon is about 50kΩ. According to Ohm's Law, when the plate current swings from its DC value of 1.2mA to a value of 0mA, the plate voltage swings by

$$(1.2mA)(50k\Omega) = 60V$$

The plate voltage therefore swings from its DC value of 108V to a value of $108V + 60V = 168V$. The AC load line therefore includes the DC operating point and the endpoint 168V, 0mA, as shown in green.

The grid can swing by 1.9V peak before the onset of severe distortion, so input headroom is 1.9V peak (+2.6dBV). Without pedal boost, a guitar pickup not likely to overdrive this stage.

The first-stage unloaded voltage gain and output impedance are the same as for the 5D3: 35.2 (30.9dB) and 20kΩ, respectively.

The unloaded signal level driving the volume and tone control circuit needs to be +5.2dBV for full power. The corresponding signal level at the guitar input jack is

$$5.2dBV - 30.9dB = -25.7dBV\ (73mV\ peak)$$

With the volume and tone controls cranked to maximum the required signal level for full power is

$$-11.7dBV - 30.9dB = -42.6dBV\ (10mV\ peak)$$

Both levels are within the capabilities of a single-coil pickup.

## Substituting a 12AX7 for the 12AY7

A common modification to the 5E3 is to substitute a 12AX7 for the original 12AY7. This is such a common practice that Fender incorporates the tube substitution into its '57 Deluxe reissue. The tube draws less DC current, which reduces the drop in voltage across the 22kΩ ripple filter resistor. Fender measures a plate supply voltage of 250V for the 12AX7 preamp and phase inverter, representing an 8-percent increase, which we will ignore for our analysis here.

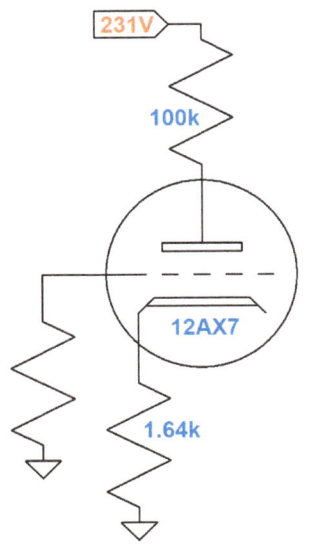

For DC the capacitors are open circuits. The 820Ω cathode resistor carries the cathode current of two triodes, so to create the same cathode voltage with only one triode its value needs to double: 1.64kΩ.

The red DC load line joins the 231V plate supply voltage on the X axis with a Y-axis value of

$$\frac{231V}{100k\Omega + 1.64k\Omega} = 2.27mA$$

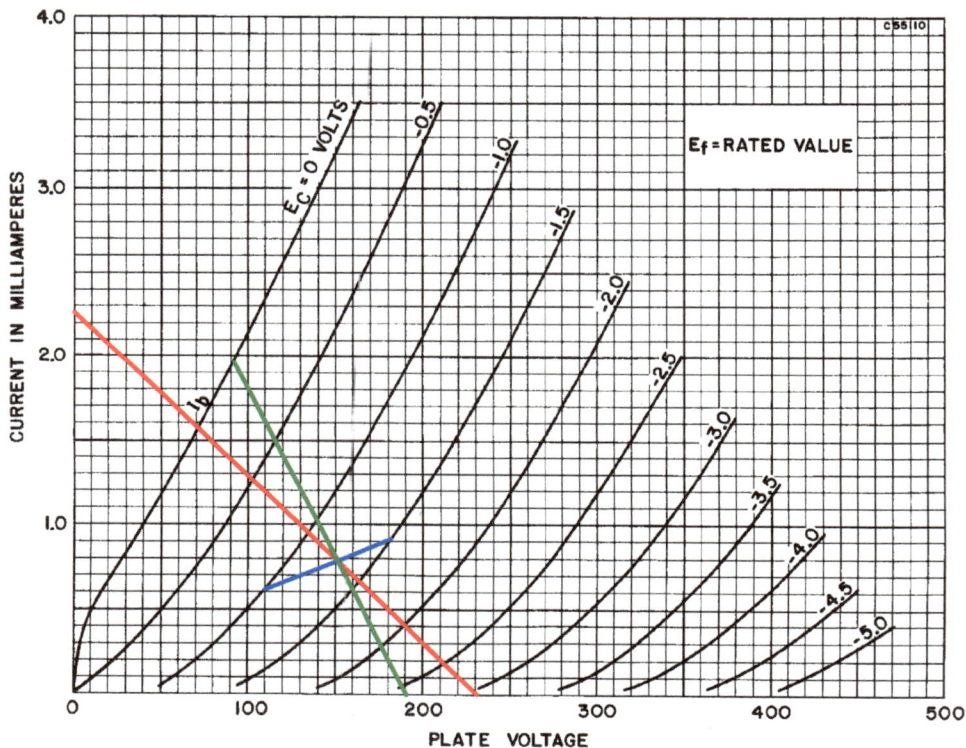

According to Ohm's Law, if the cathode voltage is -1V or -1.5V, the values for plate current passing through the 1.64kΩ cathode resistor are

$$\frac{1V}{1.64k\Omega} = 0.61mA$$

$$\frac{1.5V}{1.64k\Omega} = 0.91mA$$

These values are the endpoints for the line segment shown in blue. The DC operating point is defined by the intersection: a DC grid-to-cathode voltage of -1.3V, a DC plate voltage of 151V, and a DC plate current of 0.8mA.

**74 – CHAPTER 3**

For our worst-case AC plate load of 50kΩ, when the plate current swings from its DC value of 0.8mA to a value of 0mA, the plate voltage swings by

$$(0.8mA)(50k\Omega) = 40V$$

The plate voltage therefore swings from its DC value of 151V to a value of $151V + 40V = 191V$. The AC load line includes the DC operating point and the endpoint 191V, 0mA, as shown in green.

The triode has a maximum grid swing of 1.3V peak (-0.7dBV), representing input headroom. This is almost 10dB lower than for the 5D3's 12AY7, so the tube is driven harder. Without pedal boost, however, a guitar pickup is not likely to create 1.3V peak, so the stage is not overdriven.

For a 12AX7 amplification factor of 100, a 12AX7 plate resistance of 62.5kΩ, a plate load of 100kΩ, and a fully bypassed cathode resistor, the unloaded voltage gain from grid to plate is

$$\frac{(100)(100k\Omega)}{100k\Omega + 62.5k\Omega} = 61.5 \ (35.8dB)$$

The output impedance is equal to the plate load resistor in parallel with the plate resistance:

$$\frac{1}{\frac{1}{100k\Omega} + \frac{1}{62.5k\Omega}} = 38.5k\Omega$$

The higher output impedance increases insertion loss for the volume and tone control circuit. SPICE AC analysis demonstrates that for the bright channel with both knobs at 10-percent resistance, insertion loss is about 18.9dB at 1kHz, less for bass and slightly more for treble. This plot assumes the volume control for the normal channel is set to minimum.

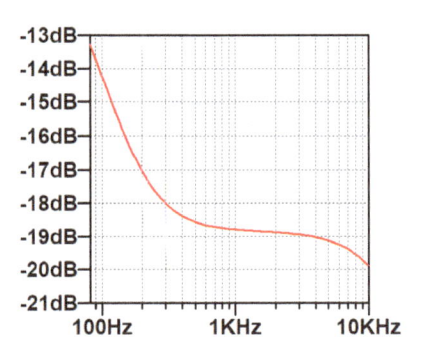

With bright channel knobs at maximum, insertion loss is about 1dB.

The second-stage grid needs a signal level of -12.2dBV to drive the power amp to full power. With the active channel's volume and tone controls at 50-percent rotation, they have 18.9dB insertion loss, which includes the first-stage output impedance. The signal level at the guitar input jack needed to drive the power amp to full power, representing input sensitivity, is

$$-12.2dBV + 18.9dB - 35.8dB = -29.1dBV \ (50mV \ peak)$$

With the instrument channel's controls at maximum, sensitivity is

$$-12.2 dBV + 1 dB - 35.8 dB = -47 dBV\ (6.3mV\ peak)$$

As a system, the amp has substantially more gain with 12AX7s in both sockets, as expected.

The 25µF cathode bypass capacitor and 0.1µF coupling capacitor are relatively large for these positions, so together they create little bass attenuation.

When shielded, the 12AX7 has 1.7pF capacitance between the grid and plate and 1.8pF capacitance between the grid and cathode. We determined that the unloaded voltage gain is 61.5 (35.8dB), so the Miller capacitance is

$$1.8pF + (61.5 + 1)(1.7pF) = 108pF$$

Using loaded gain the Miller capacitance is slightly less, so this is more capacitance than we get in reality. The series resistance in front of the grid is approximately equal to the jack's two 68kΩ resistors in parallel for an effective resistance of 34kΩ, as shown below. Ignoring the output impedance of the guitar, the -3dB break frequency is

$$\frac{1}{2\pi(34k\Omega)(108pF)} = 43kHz$$

This is much greater than guitar frequencies so the input circuit causes very little treble attenuation.

## A New, Iconic Front End

The 5E3 has a new input network connecting the jacks to the grids.

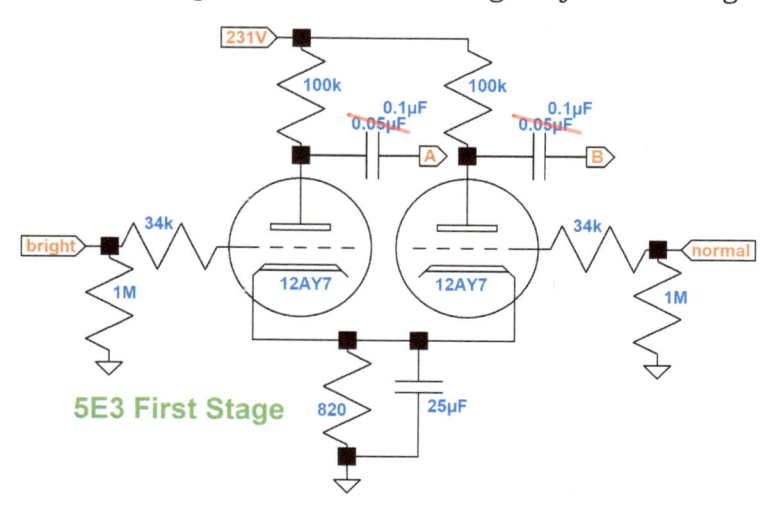

Early tweed and "woody" circuits, including the Deluxe, often use a 2-resistor input network with 50kΩ or 75kΩ values. Here, for example, is the front end of the Princeton 5C2.

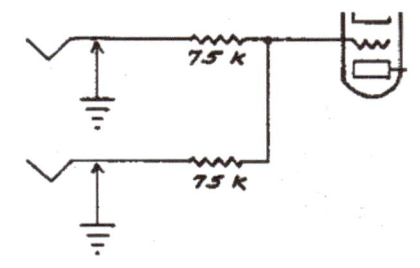

When only one instrument is plugged in, the resistors form a voltage divider that attenuates the signal by 6dB. When two instruments are in use, the amp relies on the guitar circuits to act as grid leaks to keep the grid at DC ground potential. Fender's new input network eliminates both of these concerns. It is destined become ubiquitous for guitar amplifiers.

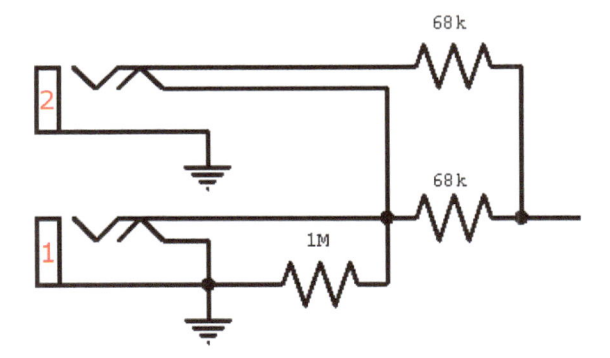

If only the No. 1 jack is used, the input circuit is a *gamma network*. The 68kΩ resistors are in parallel creating a 34kΩ grid stopper. Audio signals pass without any attenuation. The 1MΩ resistor serves as a grid leak that is independent of the guitar circuits.

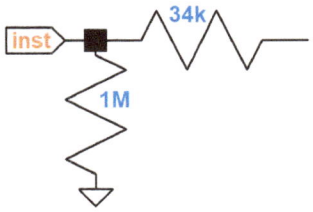

If only the No. 2 jack is used, the 68kΩ resistors form a voltage divider that attenuates the guitar signal by 6dB, just like the Princeton circuit. Today we think of the No. 1 jack as "high-sensitivity" and the No. 2 jack as "low-sensitivity."[80] The original purpose of the No. 2 jack, however, is for a second guitar, as stated by a warning tag reportedly found on a tweed Bandmaster.[81]

---

[80] Dave Hunter, **Guitar Rigs**, (San Francisco: Backbeat Books, 2005), p. 66.
[81] Tom Wheeler, **The Soul of Tone**, (Milwaukee: Hal Leonard, 2007), p. 152.

> **IMPORTANT**
> This amplifier has a built-in noise silencing circuit - Therefore it is necessary to use No. 1 jack first to realize full-volume when using No. 2 jack.

If the No. 1 jack is always in use, the circuit always provides high-sensitivity, whether for one or two guitars.

> "I believe this design came about because it was so common for people to plug more than one device into their amplifier, and they needed a variety of patching options. The Fender Museum has a '50s Twin on display that we bought from an old accordion player. Inside the amp was an old photo of the guy playing a gig, and he's got his accordion and a vocal mike both plugged into his Twin."[82] –Shane Nicholas, Fender Product Development

The new input circuit is also incorporated into the 5E Series Champ, Princeton, Super, Pro, Bandmaster, and Tremolux.

Modern amps with only one input jack often overlook the parallel origins of the grid stopper and use 68kΩ. If a 68kΩ grid stopper is used in the 5E3, the -3dB break frequency is reduced by half, but is still well above audio frequencies.

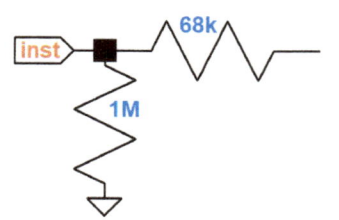

## System Profile

The schematic on the next page documents what we now know about the 5E3. System sensitivity is -25.7dBV (73mV peak). With both controls cranked to maximum, sensitivity is -42.6dBV (10mV peak). Following the schematic is a system profile for the bright channel.

The blue trace shows the signal level at the input to each stage at full power with the volume and tone controls at 50-percent rotation. The orange trace is with the controls at maximum. The yellow traces show input headroom for the preamps, phase inverter, and power amp. The gray trace shows AC ripple for the plate, screen, and triode supplies. All traces are in dBV.

---

[82] Tom Wheeler, **The Soul of Tone**, (Milwaukee: Hal Leonard, 2007), p. 49.

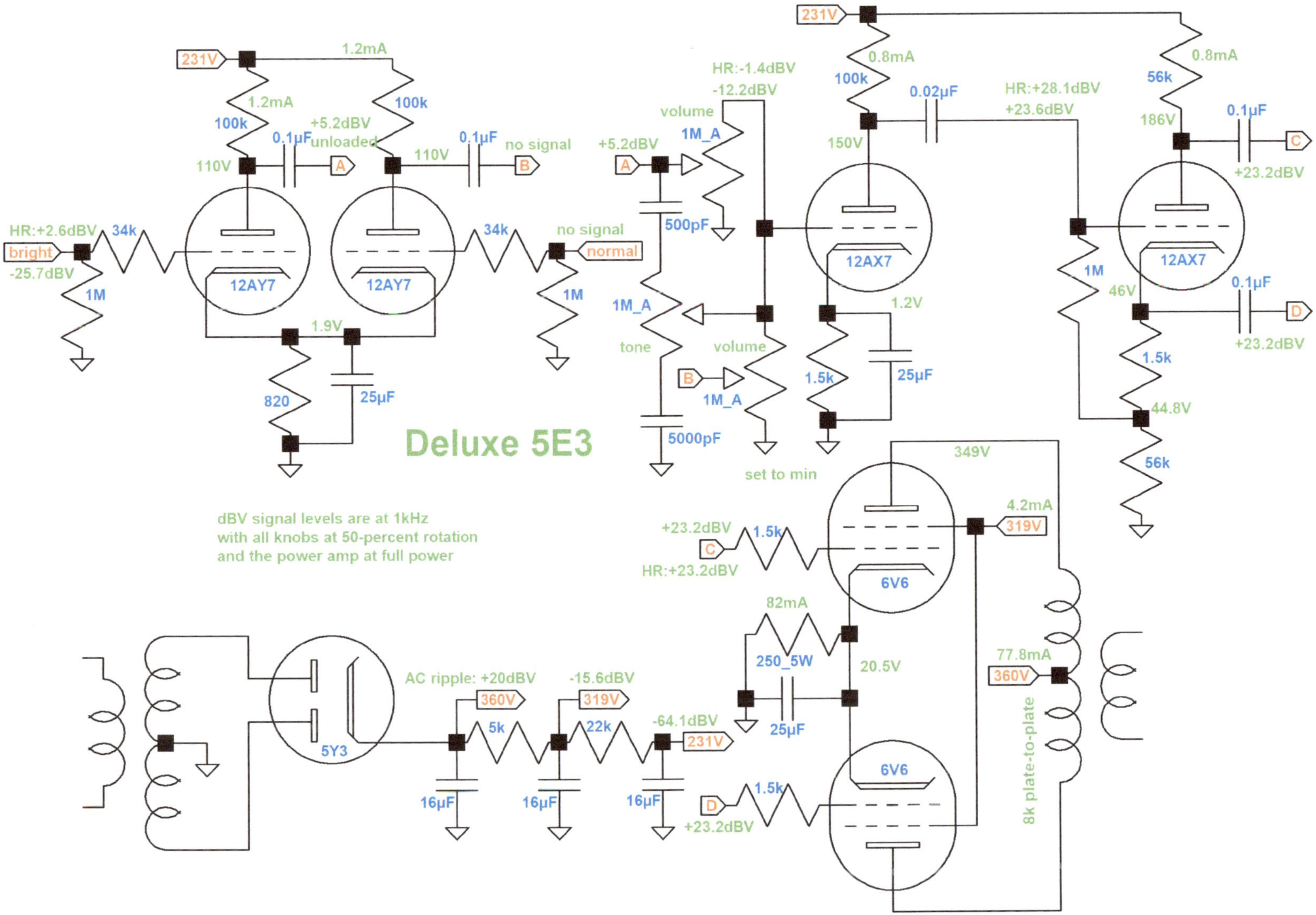
Deluxe 5E3
dBV signal levels are at 1kHz
with all knobs at 50-percent rotation
and the power amp at full power
231V
1.2mA
100k
110V
0.1µF
+5.2dBV
unloaded
no signal
HR:+2.6dBV
bright
-25.7dBV
34k
1M
12AY7
1.9V
820
25µF
normal
volume
1M_A
500pF
tone
5000pF
set to min
HR:-1.4dBV
-12.2dBV
0.8mA
150V
12AX7
1.2V
1.5k
0.02µF
HR:+28.1dBV
+23.6dBV
56k
186V
46V
44.8V
+23.2dBV
349V
4.2mA
319V
HR:+23.2dBV
6V6
82mA
250_5W
20.5V
77.8mA
360V
8k plate-to-plate
5Y3
AC ripple: +20dBV
-15.6dBV
-64.1dBV
5k
22k
16µF

## 5E3 System Profile

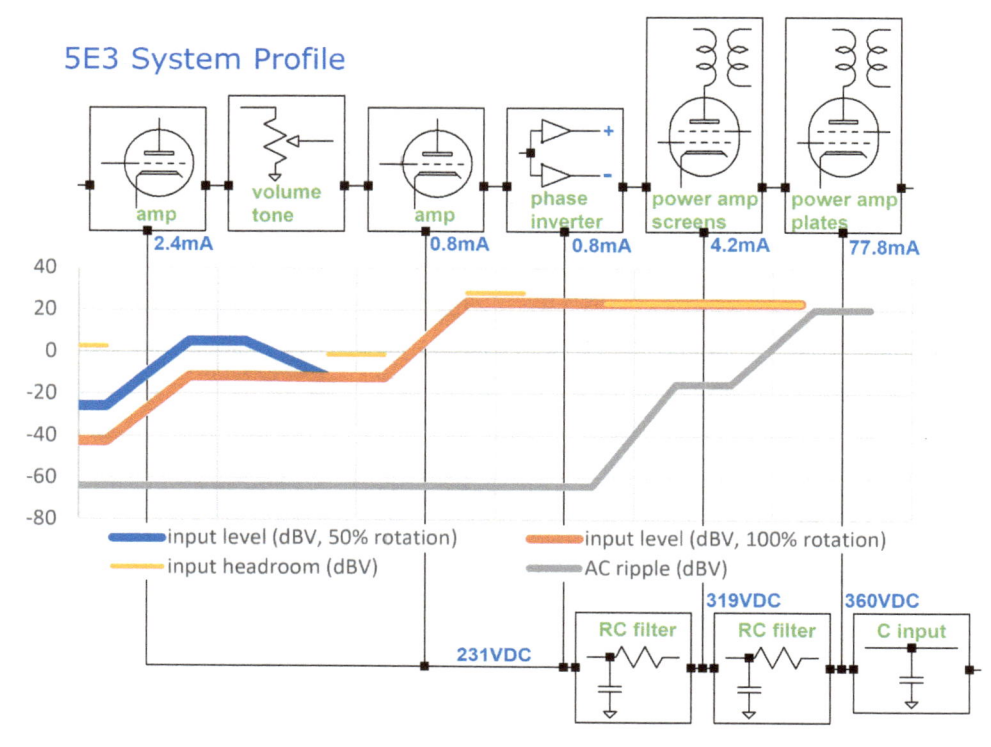

*[Figure: Block diagram of 5E3 signal chain — amp (2.4mA) → volume/tone → amp (0.8mA) → phase inverter (0.8mA) → power amp screens (4.2mA) → power amp plates (77.8mA); with power supply nodes 231VDC, 319VDC (RC filter), 360VDC (RC filter, C input). Graph plots input level (dBV, 50% rotation), input level (dBV, 100% rotation), input headroom (dBV), and AC ripple (dBV) across stages, vertical axis from -80 to 40.]*

The 5D3's 6dB attenuation at the input jack disappears, making the 5E3 more sensitive by that amount. First-stage input headroom is slightly reduced because of a lower plate supply voltage. This has no impact because it far exceeds the capability of a guitar without pedal boost. For both models the first-stage preamp is not overdriven.

## System Voicing

> "More definitively, it's a cathode-biased 15-watt amp based on a pair of 6V6GT output tubes, with no negative feedback loop around the output stage, features that together help it sound hot, juicy, and saturated at relatively low volumes, with a pronounced yet rather compressed midrange hump aided by the 6V6s."[83] –Dave Hunter

System voicing, the relative frequency response from the input jack to the power amp grids, is different for the two channels. With all knobs at noon here is the frequency response relative to 820Hz for the normal and bright channels.

---

[83] Dave Hunter, "Fender 5E3 Deluxe," **Vintage Guitar**, April 2008.

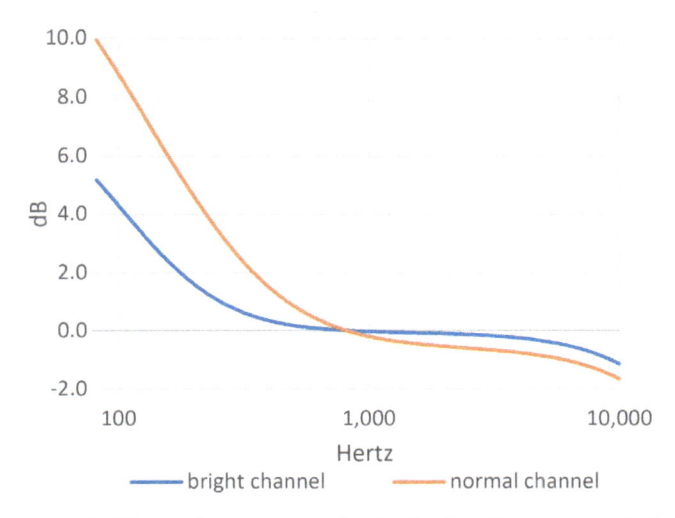

There is substantial bass boost, particularly for the normal channel.

> "The first thing that you have to know about the Torres Tweed Deluxe, or the original, is that the Normal channel has a pretty dark tonality, while the Bright channel has enough treble in it to slice through anything."[84]

Below 82Hz the response eventually falls off due mostly to coupling capacitors.

## Full Power Performance – Class A Operation?

Here is a SPICE transient simulation[85] of one cycle of a 1kHz sine wave at full power.

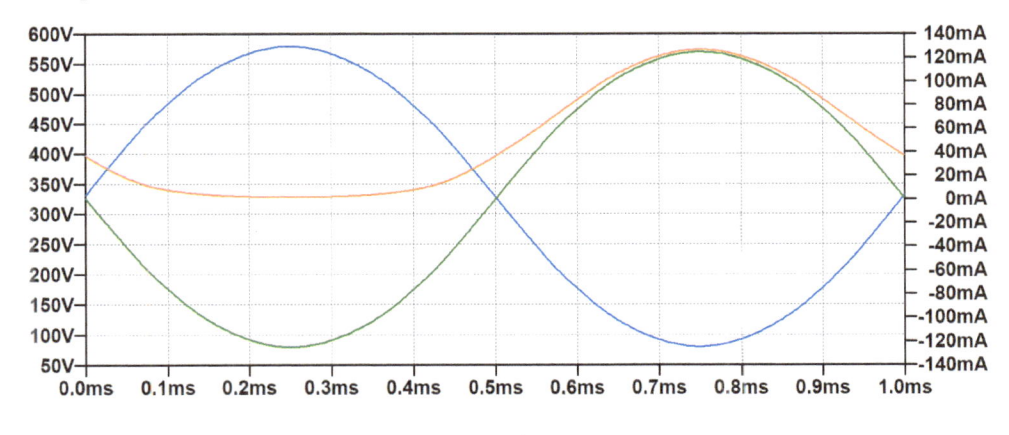

---

[84] Bob Thomas, "Torres Engineering," **Sound on Sound**, February 1, 2014. Available at https://web.archive.org/web/20140201102001/http://www.soundonsound.com/pm/nov07/articles/torrestweed.htm (Retrieved November 24, 2020)

[85] Richard Kuehnel, **Guitar Amplifier Electronics: Circuit Simulation**, (Seattle: Amp Books, 2019).

The traces show upper 6V6 plate-to-cathode voltage (blue), upper 6V6 plate current (orange), and net plate current (upper 6V6 plate current minus lower 6V6 plate current, green). Plate current and screen current average 53mA and 7mA, respectively, for each tube. Screen dissipation increases from less than 1W at idle to 2.2W at full power.

This plot shows the relationship between plate current and plate voltage plotted on estimated plate characteristics for a screen-to-cathode voltage of 299V.

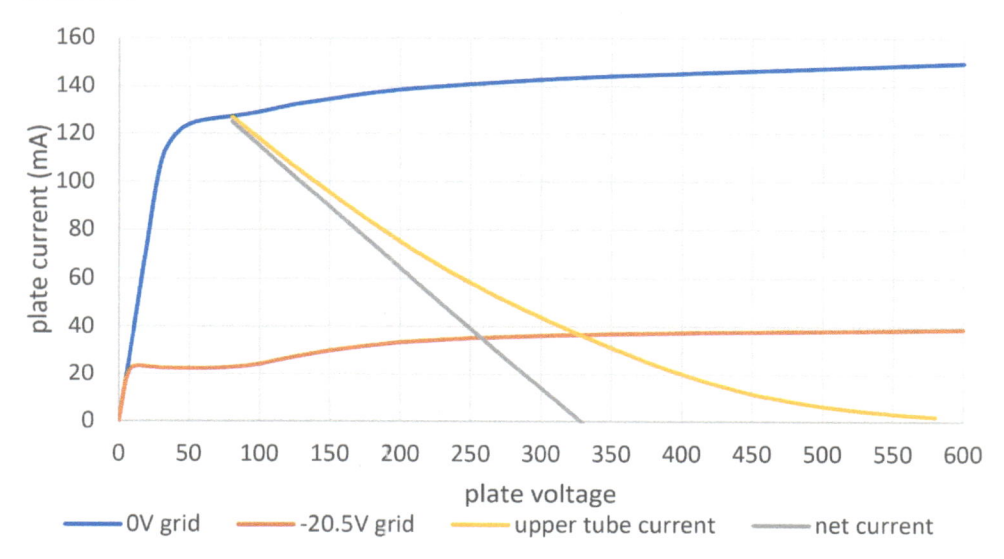

The plot assumes that supply voltages and the DC operating point hold steady at their idle values, which is approximately the case for the initial attack of a full power signal before DC voltages have had a chance to react.

The characteristic curves are for a 0V grid (blue) and a -20.5V grid (orange). The yellow trace shows plate current versus plate voltage for the upper tube. The gray curve (approximately a line) shows the net current through the transformer primary. Only positive net current is shown.

With the grid at -20.5V, the plate voltage is 328V and upper tube plate current is 38mA, corresponding to the DC operating point. When the grid voltage swings to 0V, plate voltage swings to 80V and plate current swings to 127mA. At the other extreme, with a -41V grid we get 580V, 2mA, which is slightly above cutoff, so at full power the tube conducts plate current through an entire cycle, representing Class A operation,[86] at least before

---

[86] Richard Kuehnel, **Guitar Amplifier Electronics: Basic Theory**, (Seattle: Amp Books, 2018), pp. 114-117

the DC grid bias and supply voltages have had a chance to shift the amp closer toward Class B.

> "If, in the late '50s, Fender had been thinking like a major amp maker's marketing department circa 2008, it would have billed the tweed Deluxe as a "class A amp." That's essentially what it is, as far as the common use of the term goes, which is to say it's as much of one as a Vox AC15."[87] –Dave Hunter

Bias excursion and power supply sag are discussed later in this chapter.

When the upper 6V6 grid is at 0V, the lower 6V6 grid is at -41V. At that instant 127mA flows upward through the primary while 2mA flows downward, as shown by the yellow curve, for a net upward current of 125mA, as shown by the gray curve.

When the upper grid is at -20.5V, the lower grid is also at -20.5V. There is 38mA upward and 38mA downward through the primary for a net current of 0mA. The gray trace is therefore approximately a line from 80V, 125mA to 328V, 0mA. According to Ohm's Law, the impedance corresponding to the line is

$$\frac{328V - 80V}{125mA} = 2k\Omega$$

This represents one fourth of the 8kΩ plate-to-plate transformer primary impedance. Output power is approximately

$$\left(\frac{1}{2}\right)(328V - 80V)(125mA) = 16W$$

This is before power supply voltage sag. It also assumes an ideal output transformer and an 8Ω purely resistive speaker. In the context of electronic circuits in general, this is 16 watts of power created by a relatively large voltage swing and a relatively small current swing, which is typical of a high-impedance vacuum tube circuit. The output transformer *transforms* this power to a low-impedance environment. In the secondary, voltage swings by 16V peak and current swings by 2A peak. According to Ohm's Law, this represents an impedance of

$$\frac{16V}{2A} = 8\Omega$$

which matches speaker impedance. Approximate output power is the same as in the primary:

---

[87] Dave Hunter, "Fender 5E3 Deluxe," **Vintage Guitar**, April 2008.

$$\left(\frac{1}{2}\right)(16V)(2A) = 16W$$

As the plate voltage swings down toward 80V, the highly positive screen becomes more attractive to negatively charged electrons, which increases screen current (red trace shown here) and instantaneous screen dissipation (blue trace) substantially.

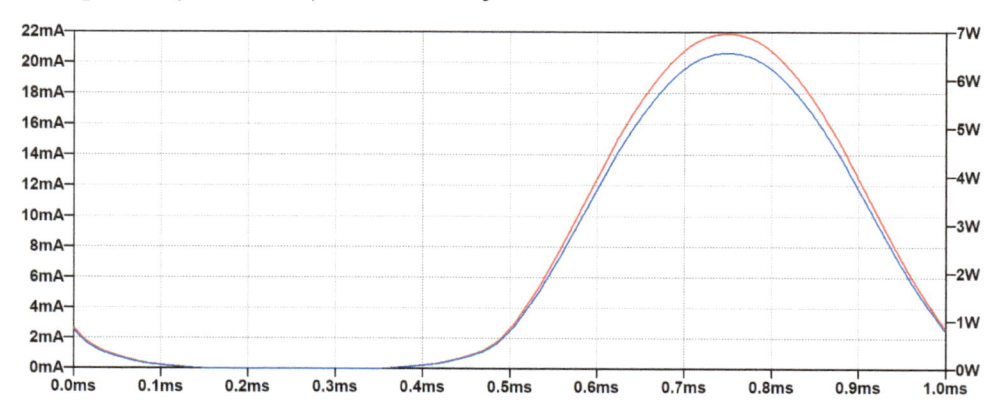

Average screen current dissipation is 2.2W, slightly more than the 2W limit of a 6V6.

## Harmonic Distortion at Full Power

Under the direction of Leo Fender, the company aims to increase power and reduce distortion, as suggested by its description of the brownface Vibrasonic.

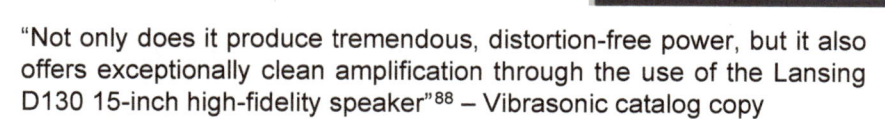

> "Not only does it produce tremendous, distortion-free power, but it also offers exceptionally clean amplification through the use of the Lansing D130 15-inch high-fidelity speaker"[88] – Vibrasonic catalog copy

At first glance, Fender achieves its objectives. SPICE Fourier analysis[89] indicates that for perfectly matched power tubes and an ideal transformer driving an 8Ω purely resistive load, the power amp produces only about 1.9-percent total harmonic distortion (THD) at full power. On the next page is a graph showing each of the components. Push-pull power amps create distortion dominated by odd harmonics, the 3rd-harmonic in particular. Even harmonics are almost non-existent assuming a perfectly balanced phase inverter (blue bars).

---

[88] Tom Wheeler, **The Soul of Tone**, (Milwaukee: Hal Leonard, 2007), p. 202.
[89] Richard Kuehnel, **Guitar Amplifier Electronics: Circuit Simulation**, (Seattle: Amp Books, 2019), pp. 139-140.

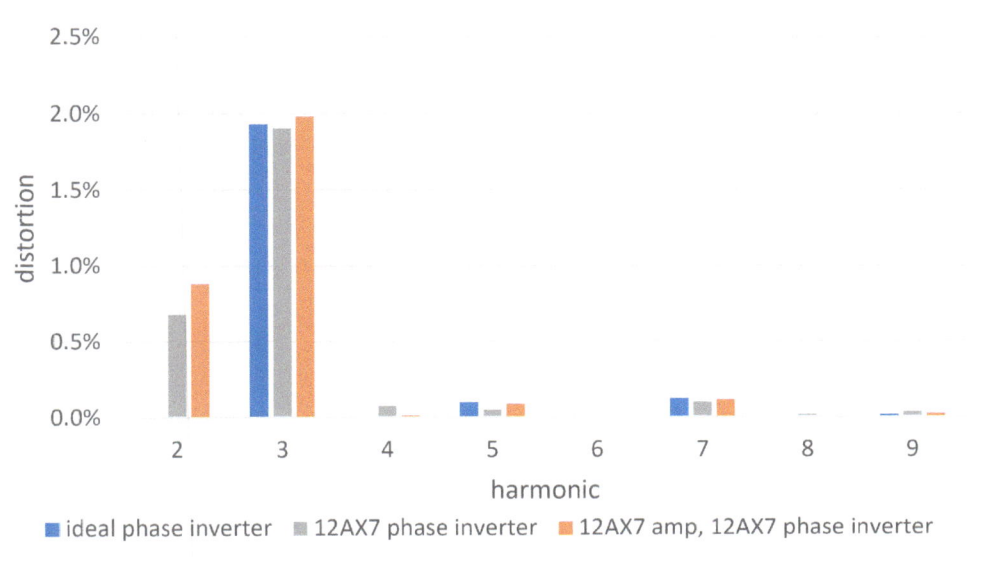

■ ideal phase inverter　■ 12AX7 phase inverter　■ 12AX7 amp, 12AX7 phase inverter

When driven by the 5E3's 12AX7 phase inverter (gray bars), THD increases slightly to 2.0 percent and $2^{nd}$-harmonic distortion increases. This assumes the phase inverter is driven by a perfect sine wave. When the phase inverter is instead driven by the 5E3's 12AX7 voltage amplifier (orange bars), THD increases to 2.2 percent and the $2^{nd}$ harmonic increases further. Both triodes are operating within their headroom limits, so their contributions to distortion are not extreme.

These SPICE simulations assume the 6V6 tubes are identical, something impossible to achieve in practice. For comparison, a 14W Class AB example from a GE 6V6GT data sheet (285V screens and plates, 8kΩ plate-to-plate primary, -19V grids, 19V peak grid voltage swing) has a measured THD of 3.5 percent, which is more realistic. Even 3.5-percent THD at full power is fairly clean. Classic college texts of the 1950s typically design for maximum power output at a specified distortion level. A level of 5-percent total harmonic distortion was considered quite acceptable at the time.

The simulation also assumes that the plate and screen supply voltages do not sag and the DC cathode voltage holds steady at 20.5V. This is also unrealistic. Average plate current for two tubes at full power is 106mA. Average screen current is 14mA. This puts 120mA through the cathode resistor, a big increase compared to idle conditions. Over time the DC grid bias becomes more negative, pushing the power amp towards Class B. The increased current load on the power supply causes supply voltages to sag, however, which reduces current through the tubes, counteracting the decrease in grid bias. These effects are explored later in this chapter.

An "8Ω" guitar amplifier speaker is far from 8Ω over much of its frequency range. Here is a graph of sound pressure and impedance versus frequency.

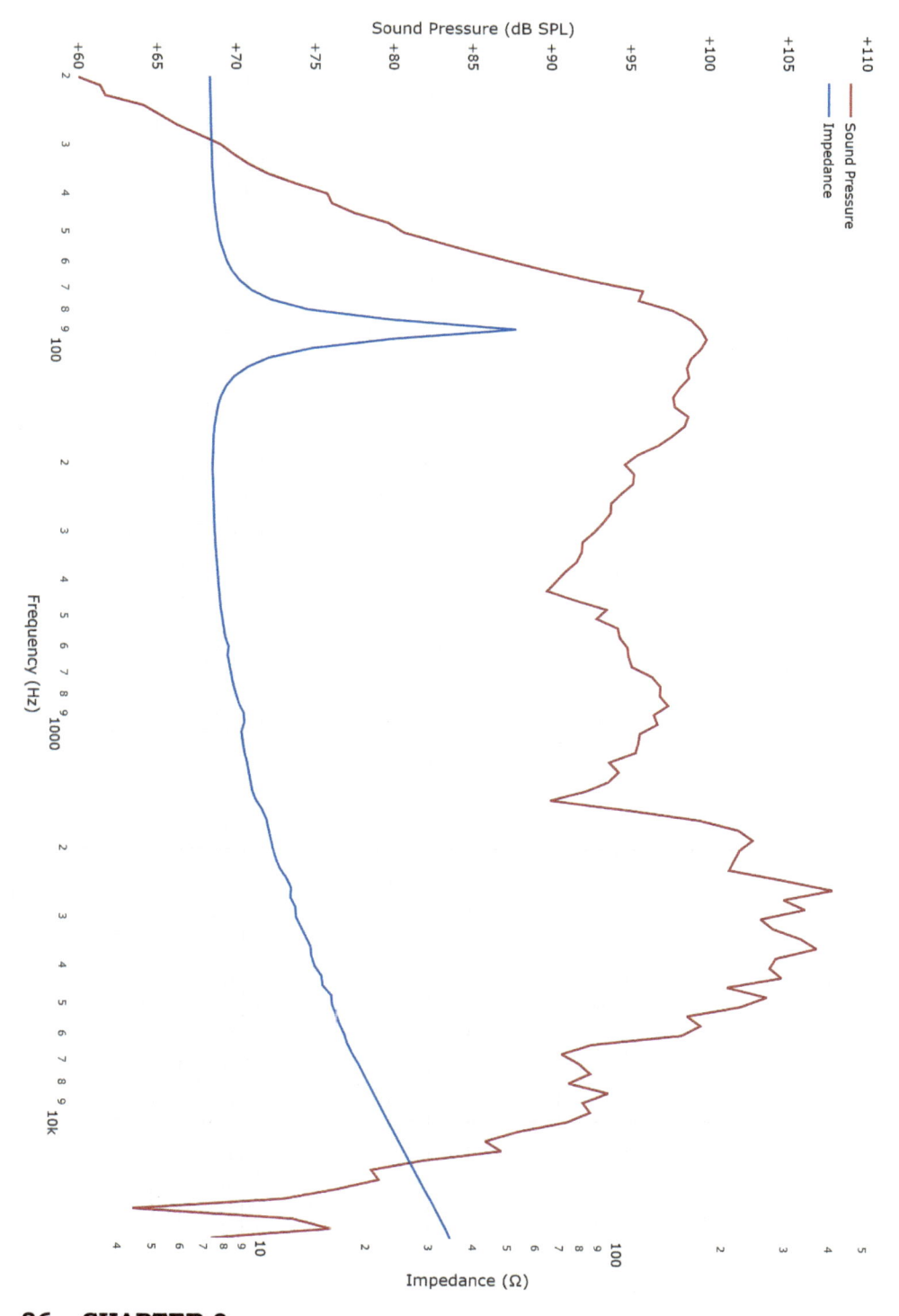

86 – CHAPTER 3

The graph, which is for a Jensen 12R-8Ω, shows features that are typical for a speaker. The impedance reaches a peak of more than 50Ω near the speaker's resonant frequency of 88Hz. The high impedance is reflected back to the primary and changes the path of the AC load line.

Moreover, a speaker is not a pure resistance – it has reactance that transforms a load "line" into more of a load "ellipse" in which different paths are taken in each direction of plate voltage swing. All speakers deviate substantially from a pure, constant-value resistance that we assume during the design process.

*Damping factor* is defined as the ratio of speaker impedance to the output impedance of the power amplifier. It is widely used as an indication of the ability of the amp to handle a frequency-dependent speaker load. Negative feedback from the output transformer secondary increases the damping factor, giving the amp more control over the speaker and a flatter response.

The following plot[90] shows typical frequency responses for a power amp connected to a real-world speaker.

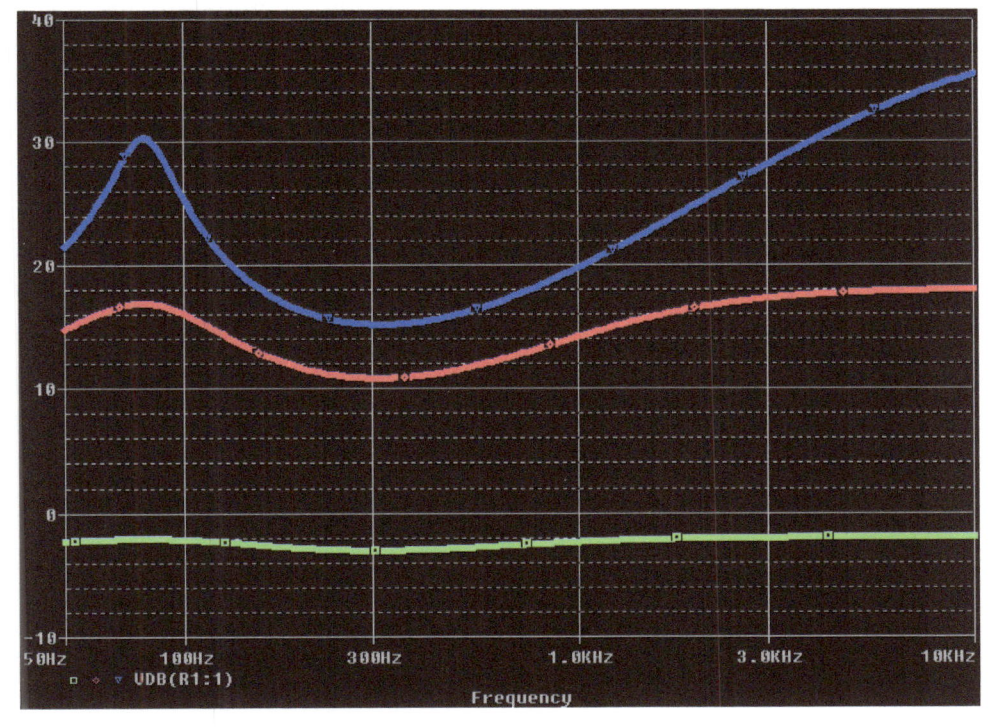

*Paul Reid*

---

[90] Personal correspondence with Paul Reid, June 2018.

The traces are for three different damping factors:

- 0.1 (upper trace), typical of a pentode without negative feedback,
- 1 (middle trace), typical for a guitar amp with feedback, and
- 10 (lower trace), typical for a high-fidelity amplifier.

The traces demonstrate that damping factor has a major impact at the low and high end of the guitar's frequency range.

At the low end, negative feedback gives the power stage more control at frequencies near speaker resonance, forcing the speaker to reproduce the power amp input signal instead of following its flabby, booming whims. This gives negative feedback a reputation for tightening up the low end. A cathode-biased amp without feedback has grunge, dirt and what Dave Hunter describes as "a little gainier, looser, and more saturated sounding."[91]

At high frequencies, negative feedback reins in the treble response by reducing high-frequency distortion and flattening the speaker's inherent high-frequency boost. The response is similar to an amp with feedback and its presence control set to minimum. The Deluxe, in effect, has a presence control that is permanently cranked to maximum, creating warmth, sparkle, and what Shane Nicholas at Fender Product Development describes as "sizzle."[92]

## Intermodulation Distortion (IMD)

To measure harmonic distortion, the input is a perfect sine wave at a specified fundamental frequency, representing a laboratory measurement in its most basic form. The last section, for example, uses a 1kHz sine wave of sufficient amplitude to drive the power amp to full power and then measures the harmonic frequencies that are multiples of the input: 2kHz, 3kHz, 4kHz, etc. Moreover, the measurements are made using many simplifying assumptions, most importantly

1. a perfect sine wave input,
2. identical power tubes,
3. an ideal output transformer,
4. an 8Ω purely resistive speaker,
5. no power supply voltage sag, and
6. no power amp bias excursion.

For a real guitar amp, *intermodulation distortion* contains frequencies that are not multiples of the fundamental. Measuring it still has a laboratory vibe, but instead of only a single sine wave input, more than one frequency

---

[91] Dave Hunter, **Guitar Amp Handbook**, (Milwaukee: Backbeat Books, 2005), p. 67.
[92] Tom Wheeler, **The Soul of Tone**, (Milwaukee: Hal Leonard, 2007), p. 178.

is applied to the input at the same time. This represents just one tiny step toward analyzing an amp's response in a way that is closer to a real musical performance.

This SPICE transient simulation,[93] for example, uses the sum of two sine waves applied to the power tube grids: 82Hz (low E) and 123Hz (the B above low E). The signal shown here is measured at the grid of one 6V6. It combines the root and fifth of a power chord in its purest form. The sine wave amplitudes are equal and sufficient to drive the grids to their full extent of voltage swing without overdriving them.

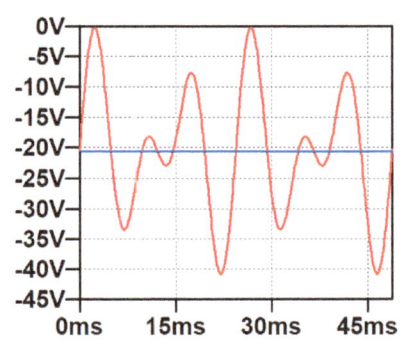

If a circuit is perfectly linear, it can change a signal's amplitude and phase but it cannot create frequencies that are not present in the input. If 82Hz and 123Hz are the only frequencies in the input, they are the only frequencies in the output. Nonlinear distortion adds new frequencies, as we observed for harmonic distortion. Here is our power chord frequency spectrum measured at the speaker assuming matched power tubes, an ideal transformer, and an 8Ω purely resistive speaker.

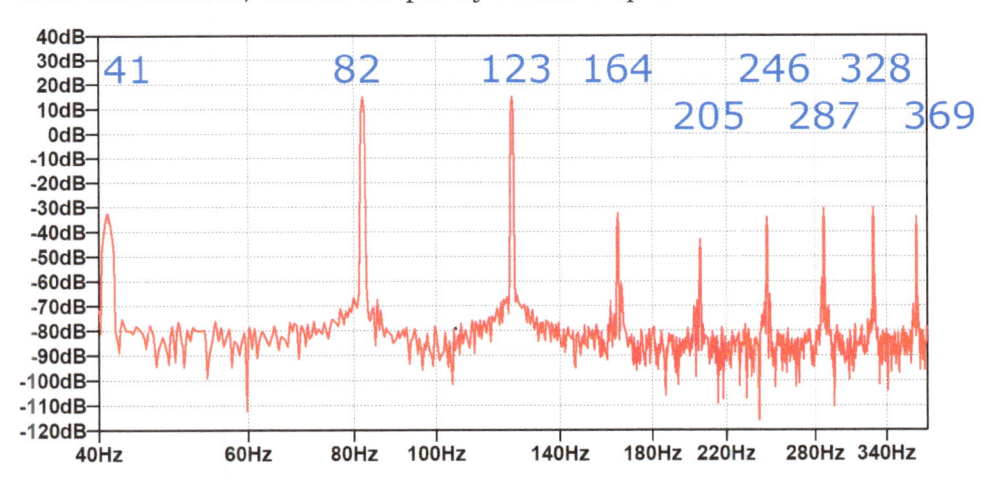

The biggest contributors to the frequency spectrum are the two fundamental frequencies: $f_1 = 82Hz$ and $f_2 = 123Hz$. The other peaks represent harmonics (multiples of a fundamental) or IMD (sums and differences of the two fundamentals).

---

[93] Richard Kuehnel, **Guitar Amplifier Electronics: Circuit Simulation**, (Seattle: Amp Books, 2019).

| Hertz | Component | Type |
|---|---|---|
| 41 | $f_2 - f_1$ | Intermodulation |
| 82 | $f_1$ | Fundamental |
| 123 | $f_2$ | Fundamental |
| 164 | $2f_2 - f_1$ | Intermodulation |
| 205 | $f_1 + f_2$ | Intermodulation |
| 246 | $3f_1$ | Harmonic |
| 287 | $2f_1 + f_2$ | Intermodulation |
| 328 | $2f_2 + f_1$ | Intermodulation |
| 369 | $3f_2$ | Harmonic |

There are more harmonics and IMD products higher than 369Hz that are not included in the plot. We observe that, unlike a single-ended power amp, a perfectly balanced push-pull design creates no 2nd harmonics.

The first IMD component represents the difference between the frequencies at the input:

$$123Hz - 82Hz = 41Hz$$

This has musical consequences. When the input contains the root and the fifth of a power chord, the fifth has a fundamental frequency that is 1.5 times the frequency of the root. The difference is a new frequency that is 0.5 times the frequency of the root, i.e. one octave lower. The amp thickens the guitar signal by adding this tone.

The input signal frequencies still dominate the output despite harmonic and intermodulation distortion. This is a good thing, because in the midst of all the tonal embellishment added by the amplifier, the human ear still perceives the result to be the root and fifth of a power chord.

## Overdrive Dynamics

> "The 'best' Tweed Deluxe foible of all is that as soon as you turn the volume up past 3 or 4, you'll find yourself running into distortion, which is why this little amp is so loved and revered. If you are looking for clean sounds then the Tweed Deluxe isn't the amplifier for you. In any event, unless you know how best to drive a Tweed Deluxe you won't get the best out of it, and you will never understand why it is worshipped in the way that it is."[94] –Bob Thomas

---

[94] Bob Thomas, "Torres Engineering", **Sound on Sound**, February 1, 2014. Available at https://web.archive.org/web/20140201102001/http://www.soundonsound.com/pm/nov07/articles/torrestweed.htm (Retrieved November 24, 2020)

## No Feedback, Maximum Grunge

Overdriving the power amp causes output clipping, which represents a lack of output response to a changing input signal voltage. The output provides the source voltage for any negative feedback, so these effects reduce it.

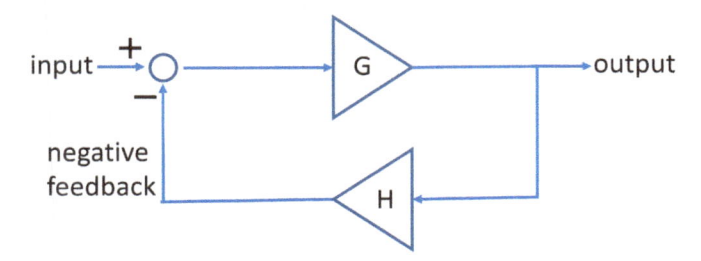

This creates more gain, which drives the amp further into an overdriven state, producing even more clipping. The net result: as the amplifier is overdriven, negative feedback from the output transformer accelerates the transition to a more distorted state. Moreover, feedback increases the contrast between clean and overdrive. At less than full power, feedback reduces distortion to create a cleaner sound. During overdrive feedback is suppressed and the power amp snarls without constraint. The net result is a snappier transition between two more distinct states.

Fender adds negative feedback to the Deluxe 6G3, the subject of the next chapter. The 5E3's lack of feedback gives it a more sustained, less snappy transition into overdrive compared to subsequent models – a tweedy growl instead of a blackface bark. With no feedback effects to kickstart distortion beyond full power, the transition from clean to saturated is gradual and dilated. Moreover, the amp is grittier at less than full power – the contrast between clean and overdriven tones is reduced. Without feedback the amp elastically transitions between two less distinct states, giving it a bluesy vibe with sponginess, compression, and sustain that are essential qualities of a tweed Deluxe.

> "Two easily overdriven cathode-bias 6V6 output tubes deliver a sweet, harmonically rich tone, and the 5Y3 tube rectifier has the sag required for dynamics and touch sensitivity. This holy grail of vintage combos has been used by Neil Young, Mike Campbell, Rich Robinson, Mark Knopfler, Billy Gibbons, and countless others."[95]

## Phase Inverter Effects

The 5E3 is the only Deluxe model with a split-load phase inverter.

---

[95] Damian Fanelli, Christopher Scapelliti and Tom Gilbert, "The 10 Most Iconic Guitar Amps of All Time," **Guitar World**, April 24, 2020.

**5E3 Voltage Amplifier and Phase Inverter**

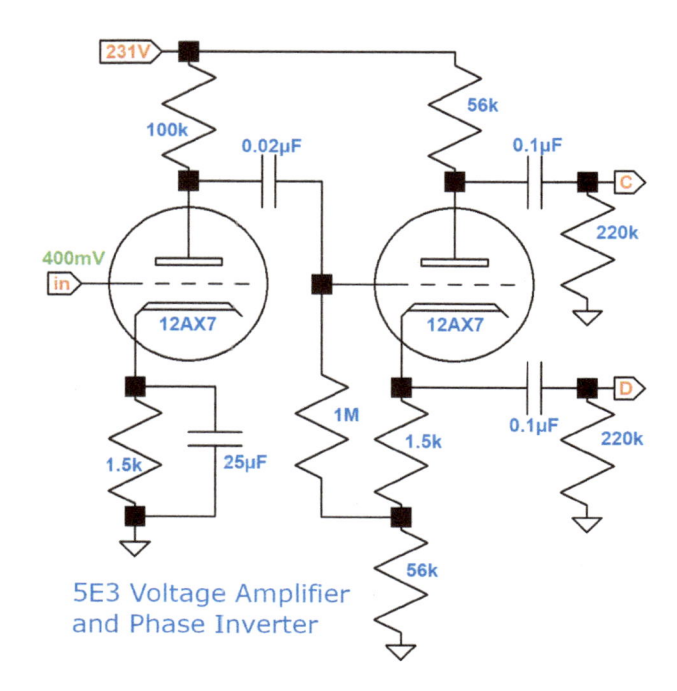

"The split-load PI adds its own distortion to the brew, the result being a warm, fuzzy mélange of tube overdrive that can get super-compressed and feel delightfully tactile – the perfect raw, low-volume vintage electric blues tone, some might say. Delightful in some circumstances, it can mush out entirely too quickly for some playing styles and doesn't quite hold together enough for well-defined twang tones or speedy picking at any kind of volume, for example."[96] –Dave Hunter

Using a SPICE transient simulation, let's drive the 12AX7 voltage amplifier upstream of the phase inverter with a 400mV peak, 1kHz sine wave and take a complete tour of the power amp and phase inverter, starting with the lower 6V6 driven by node D. To make the effects easier to interpret, we will bias the power amp at exactly -20V.

If we were to disconnect the phase inverter from the power amp grids it would be operating well within its linear range and its outputs would swing by about 22V peak. When connected, however, the signal exceeds the power amp's input headroom by 2V.

Here on the left is the grid-to-cathode voltage for the lower 6V6, the one driven by the 12AX7 cathode. There is a flattening as the voltage tries to swing positive from 0V due to 6V6 grid current, as shown on the right.

---

[96] Dave Hunter, "Fender 5E3 Deluxe," **Vintage Guitar**, April 2008.

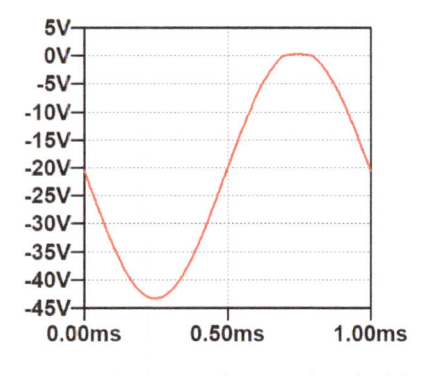

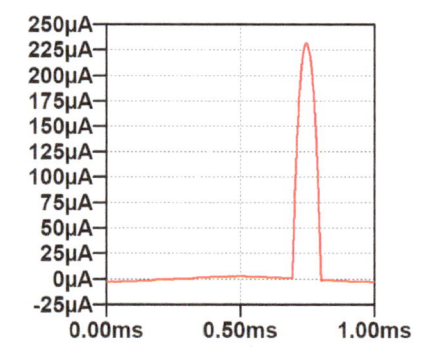

The phase inverter is overloaded by the sudden demand for current.

The 6V6 grid-to-ground voltage is 20V higher than its grid-to-cathode voltage.

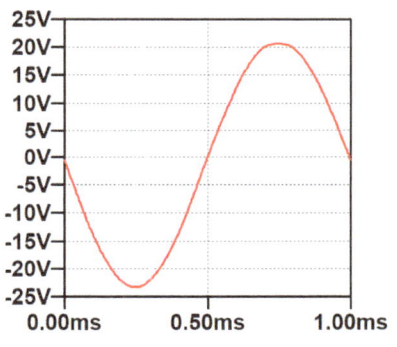

Flattening occurs as the grid reaches 20V. The phase inverter's DC cathode voltage is 46V, which is blocked by a 0.1µF coupling capacitor, so flattening of the cathode voltage occurs 46V higher, i.e. at 66V. Under these conditions, the cathode is said to be *clamped* at 66V.

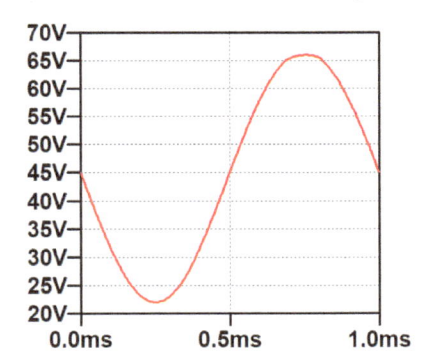

Meanwhile, the phase inverter grid voltage continues to increase unabated, so the phase inverter's grid-to-cathode voltage, i.e. the input signal perceived by the tube, gets a big boost, as shown here.

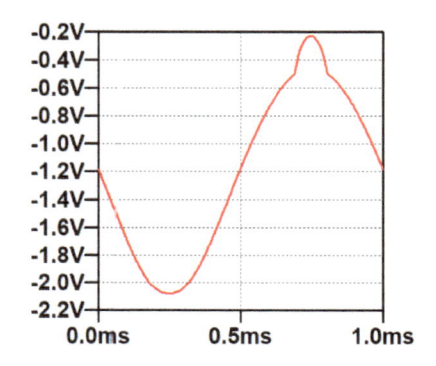

For all of the first half cycle and most of the second half cycle, the 12AX7 cathode voltage *follows* the grid voltage, creating negative feedback from cathode degeneration, just like a *cathode follower* circuit.[97] Because of this negative feedback, a split-load phase inverter normally has unity gain – it provides no voltage amplification. When the cathode is clamped, however, negative feedback disappears. The cathode voltage holds steady, just like it would for typical guitar amplifier preamp stage with a fully bypassed cathode resistor, so the grid-to-cathode voltage gets a big boost of gain.

The upper output is taken from the plate, which is inverted compared to the grid. When the cathode is clamped, the plate voltage gets a boost of amplification.

The voltage seen at the upper 6V6 grid (node C) is the same, but the triode's 186V DC plate voltage is blocked by the coupling capacitor. The upper 6V6 grid swings positive and negative relative to its 0VDC value.

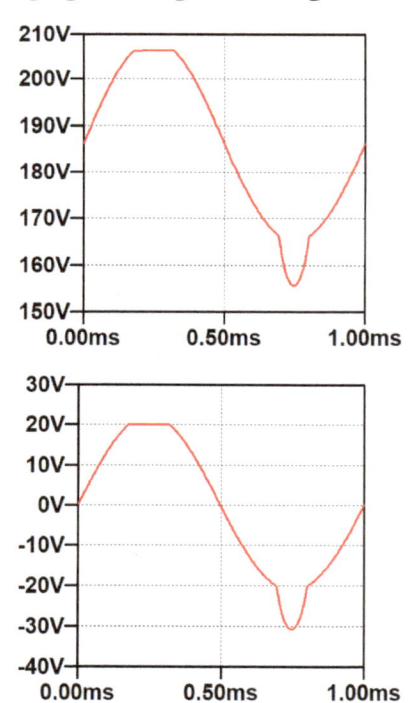

---

[97] Richard Kuehnel, **Guitar Amplifier Electronics: Basic Theory**, (Seattle: Amp Books, 2018), pp. 79-87.

Let's complete our tour of the circuit by looking at the upper 6V6 grid-to-cathode voltage. This represents the input voltage perceived by the power tube, which is 20V less than the grid-to-ground voltage. The plot shows that the big bulge due to amplification occurs as the upper tube is near cutoff. Clamping merely sends the tube into cutoff more quickly. Moreover, this assumes that the power amp's -20V DC grid bias, plate supply voltage, and screen supply voltage stay steady during overdrive. They don't! Overdrive affects all three, pushing the amp towards Class B.

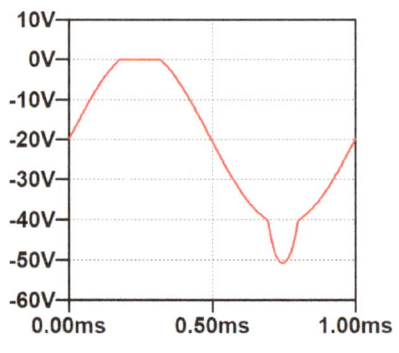

Here is a plot of the voltages at the power tube grids and the speaker.

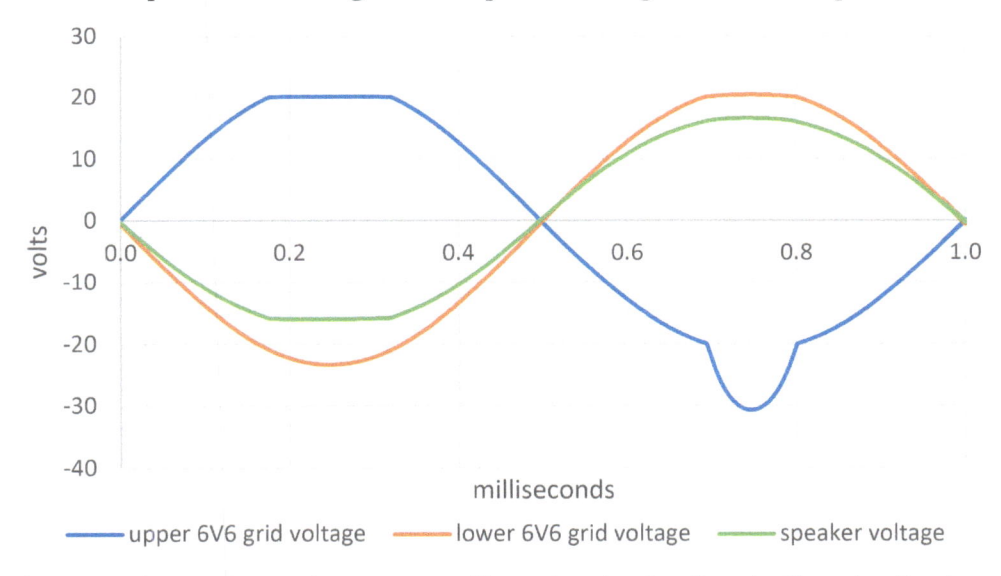

The speaker voltage shows more distortion in the first half cycle. During the second half cycle, flattening in the peak of lower 6V6 grid voltage is partially counteracted by clamping effects in the upper 6V6 grid voltage. Since the upper 6V6 is close to cutoff, those effects are minimal and completely disappear as the power amp transitions toward Class B operation.

We can conclude that while clamping of the split-load phase inverter's cathode is an interesting phenomenon, it may not have as significant an impact on the overdrive performance of the 5E3 power amp as one might suspect. On the other hand, our tour clearly shows the unique characteristics of a split-load phase inverter in overdrive. Moreover, we are once again looking at laboratory conditions – a guitar cannot generate a steady 22V 1kHz sine wave. Most importantly, the human ear is not part of our SPICE simulation

## Bias Excursion and Power Supply Sag

Here is the power amp under idle conditions. In overdrive, 6V6 grid current causes the upstream 0.1µF coupling capacitors to charge over the course of a few milliseconds. This makes the effective 6V6 DC grid bias more negative, creating bias excursion. For the power tubes, average plate and screen current increases, which eventually increases the current through the 250Ω cathode resistor, a process that is slowed by the 25µF cathode bypass capacitor. Over the course of about 10 milliseconds the cathode voltage increases substantially, making the effective DC grid bias more negative.

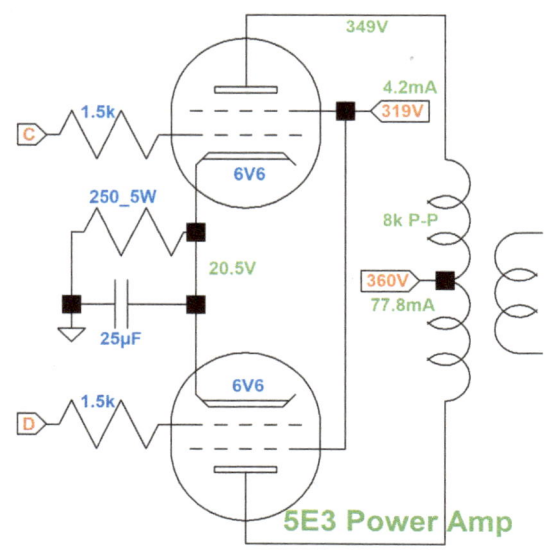

Bias excursion due to the charging of the coupling capacitors and increased cathode current pushes the power amp away from Class A and toward Class B. At its extremes, bias excursion can result in *blocking distortion*,[98] when the power tubes are in cutoff for an entire cycle of the input signal. Neil Young drives his 5E3 right up to that edge.

> "What the Deluxe offers Neil is the fine line between clean, dirt, and blocking. Some of the "one note solos" have distinct note-tones via fine control of level (and other parameters). There are passages where he is bouncing along just in/out of blocking distortion: enough to flat the wave but not enough to duck it. Other guitarists do all these things, but also other techniques. Amp abuse is the core of Neil's louder work."[99] –Paul Reid

On the next page is the power supply under idle conditions. Downstream of the 5Y3 there is a total load current of 86mA. Power supply DC output voltage is 360V, as marked by the red dot on the operation characteristics of a 5Y3.

---

[98] Richard Kuehnel, **Guitar Amplifier Electronics: Basic Theory**, (Seattle: Amp Books, 2018), pp. 166-170.
[99] Personal correspondence with Paul Reid, February 2021.

**5E3 Power Supply**

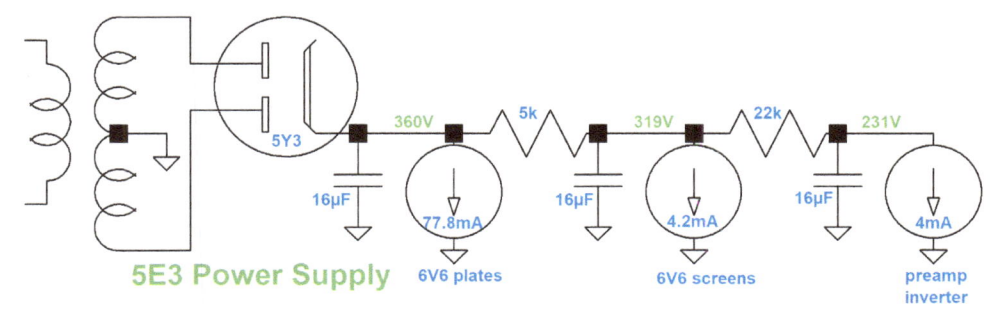

- 360V — 6V6 plates — 77.8mA — 16μF
- 5k
- 319V — 6V6 screens — 4.2mA — 16μF
- 22k
- 231V — preamp inverter — 4mA — 16μF
- 5Y3 rectifier, 16μF input cap

**OPERATION CHARACTERISTICS**
FULL-WAVE RECTIFIER WITH CAPACITOR-INPUT FILTER

$E_f$ = RATED VALUE
$C$ = 20 μf
$R_s$ = 50 OHMS FOR CURVES 1-5
$R_s$ = 135 OHMS FOR CURVES 6-8
BOUNDARY LINE DEA IS SAME AS SHOWN ON RATING CHART 1

At full power, 6V6 plate current for two tubes is 106mA. Screen current is 14mA. Total power supply load current is 124mA, so the plate supply sags by about 30V, as shown by the blue dot. This assumes, however, that average plate current and screen current hold steady over the course of the sag. In reality, full power load current decreases in response to lower supply voltages, so our dots represent boundaries for the power supply's operating range. Between these limits, the power supply executes a complex choreography in response to a real musical performance, particularly in overdrive.

**PINNACLE – THE NARROW PANEL DELUXE 5E3 – 97**

## A Pinnacle and a Pivot

Many would say the Deluxe design reaches its peak with the 5E3, becoming the ultimate tone machine for public performance and recording.

> "By the time the band moved to L.A., found new members, and became the Heartbreakers, Campbell was using mostly a Fender Broadcaster he plugged into a tweed Deluxe they found tucked away in a club, dusty and non-functional. They sprang to get it working and used it to record that first album, with "Breakdown," "American Girl" and "I Need to Know." Today, vintage tweed Deluxes remain his preferred taste onstage and in the studio."[100]

The 5E3 Deluxe has just three knobs and no superfluous features to clutter the signal path. It rejects out of hand the benefits of a feedback circuit or a fixed bias supply. It eschews the complexity of added features and creates a guitar amp in its purest form.

> "Simplicity is beauty."[101] –Victor Mason, Mojave Ampworks

Fender's minimalist design strategy for the Deluxe is about to change with the next model iteration – with an abundance of added features and a major power-amp overhaul, the Deluxe pivots in an entirely new direction. The risk pays off. Fender's perennial desire to push the design forward will win the accolades of generations of working musicians and create the ultimate desert-island amp.

---

[100] Ward Meeker, "Mike Campbell Still Thrilled," **Vintage Guitar**, July 2017.
[101] Dave Hunter, **The Guitar Amp Handbook**, (Milwaukee: Backbeat Books, 2015), p. 255.

# Chapter 4: Enhancement – The Brownface Deluxe 6G3

> "Brown Fender amps are generally cleaner sounding than tweeds at comparable volume settings, and they possess a more subtle attack with less high frequency emphasis and power than the blackface amps ... A few knowledgeable cyber-surfing amp gurus have even begun to call them the best sounding amps Fender ever made."[102]
> –Mike Letts, ToneQuest Report

The modifications introduced in the Deluxe 6G3 represent almost a complete overhaul of the existing design. The changes reflect a wave of enhancement that affects most of Fender's product line. This begs the question, can a pinnacle of achievement like the Deluxe 5E3 really be "enhanced?"

> "Leo Fender had no interest in building 'vintage' products. The quality, styling, and gold-standard tones of old Fender amps – all enhanced by time and nostalgia – may have imbued them with a vintage radiance and rendered them icons of a bygone era, but to Leo Fender it was all about the new. He kept his eye on current trends in audio and pushed his suppliers to upgrade their standards to meet his own. He strove to use improved components and the latest circuitries in order to provide players with ever more sophisticated tools."[103] –Tom Wheeler

## System Design Concept

The DC voltages shown in the schematic on the next page are Fender's measurements.

Fender's transition from tweed to tolex is marked by the addition of new features. Technically, none can be classified as an "invention," at least by the criteria of the U.S. Patent Office, but each represents a true enhancement to the existing circuit.

> "Leo's real genius was in collecting information, combining it with his experience, and blending that into a workable product that was not only fair in price but very reliable."[104] –Hartley Peavey

---

[102] Tom Wheeler, **The Soul of Tone**, (Milwaukee: Hal Leonard, 2007), p. 229.
[103] Tom Wheeler, p. 111.
[104] Tom Wheeler, p. 207.

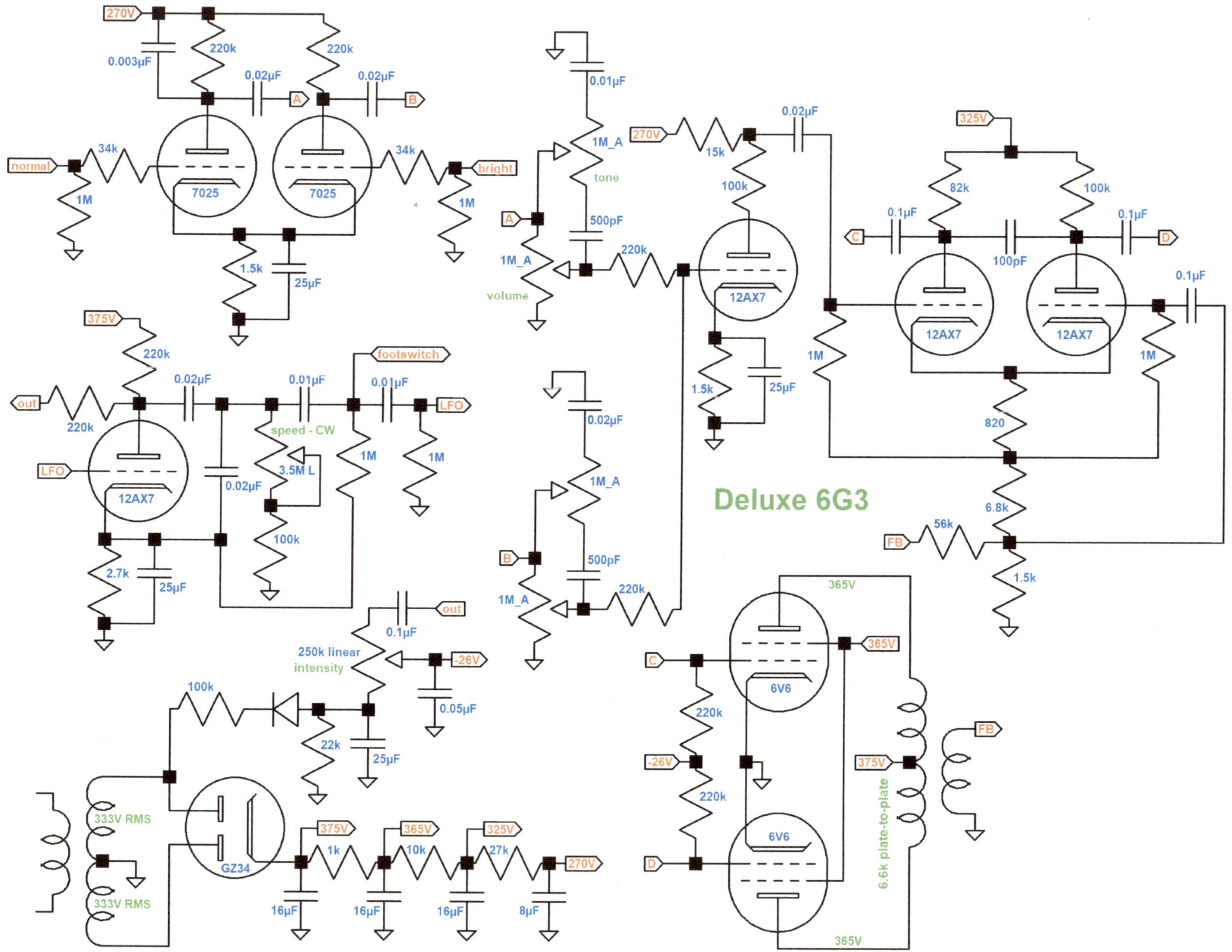
Deluxe 6G3
270V
0.003µF
220k
220k
0.02µF
0.02µF
A
B
normal
34k
34k
bright
1M
1M
7025
7025
1.5k
25µF
375V
220k
out
220k
0.02µF
0.01µF
0.01µF
footswitch
LFO
LFO
12AX7
speed - CW
3.5M L
1M
1M
0.02µF
100k
2.7k
25µF
0.1µF
out
250k linear
intensity
-26V
0.05µF
100k
22k
25µF
333V RMS
333V RMS
GZ34
375V
365V
325V
270V
1k
10k
27k
16µF
16µF
16µF
8µF
0.01µF
1M_A
tone
500pF
1M_A
volume
220k
0.02µF
1M_A
500pF
1M_A
220k
270V
15k
0.02µF
100k
12AX7
1.5k
25µF
325V
82k
100k
0.1µF
0.1µF
C
D
100pF
12AX7
12AX7
0.1µF
1M
1M
820
6.8k
56k
FB
1.5k
365V
365V
C
D
6V6
6V6
220k
220k
-26V
375V
FB
6.6k plate-to-plate
365V

One new feature for the 6G3 is tremolo, which was available in the late 1940s on amps from Danelectro, Multivox Premier, and Gibson.[105] Fender's first amp with onboard effects was the 1955 Tremolux.[106] The feature was incorporated into Fender's "Professional" series amps (the Super, Bandmaster, Pro, and Twin) in 1960.[107]

For more speaker control and less "harmonic chaos"[108] at higher power levels, the 6G3 gets some of the features already incorporated into its high-end cousins, including a GZ34 rectifier, fixed bias, and global negative feedback. Conspicuously absent from the list is an expensive choke for the power supply – the Deluxe waits a bit longer to receive this upgrade. Power supply voltages continue their upward trajectory in the new design, pushing the power tubes ever harder.

Here is a long list of upgrades for the 6G3.

**tremolo** — Tremolo is added. It uses only half of a 12AX7 dual triode, making the other triode available for an additional voltage amplification stage.

**rectifier** — A stiff GZ34 replaces the spongy 5Y3 for less DC power supply voltage sag.

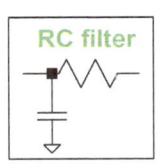

**RC filter** — 6V6 screen voltages increase by adjusting ripple filter resistor values. A third RC ripple filter is added between the phase inverter and the preamp to support an additional stage of voltage amplification.

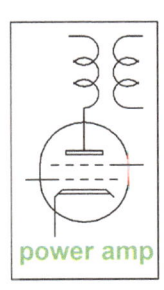

**power amp** — The power amp gets fixed bias. Higher 6V6 screen voltages and the removal of cathode bias translate to greater output power.

---

[105] John Teagle and John Sprung, **Fender Amps: The First Fifty Years**, (Milwaukee: Hal Leonard, 1995), p. 28.
[106] Tom Wheeler, **The Soul of Tone**, (Milwaukee: Hal Leonard, 2007), p. 122.
[107] John Teagle and John Sprung, p. 30.
[108] Tom Wheeler, p. 178.

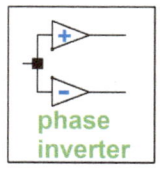

**phase inverter** — First seen in the 5F series Bassman and Twin, a long-tailed pair replaces the 5E3 split-load phase inverter. This design change is incorporated into all of Fender's 6G series push-pull models.

**feedback** — Negative feedback is added between the output transformer secondary and the long-tailed pair.

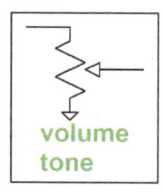

**volume tone** — The normal and bright channels get separate tone controls. The volume and tone control circuit also becomes more aggressive, with increased insertion loss. The controls are much less interactive.

**preamp** — For increased gain at the front end, a 7025 (low-noise, low-hum 12AX7) with a 220kΩ plate load resistor replaces the 12AY7 and 100kΩ resistor.

## Coaxing Even More Power from the Power Amp

A number of changes are incorporated into the power amp and power supply to increase the screen-to-cathode voltage:

1. A GZ34 rectifier replaces the 5Y3. The GZ34's lower plate resistance introduces less power supply voltage sag than the 5Y3.
2. Fixed bias replaces cathode bias. The screen-to-cathode voltage equals the full screen-to-ground voltage without a voltage drop across a cathode resistor.
3. The first RC ripple filter resistor value is reduced to 1kΩ. The voltage drop between the 6V6 plate and screen supplies is reduced to only 10V.

## Power Amp Performance

The power amp plates and screens are at the same DC voltage: 365V. This means we can use the data sheet's triode-connected plate characteristics to determine the DC cathode current, i.e. the sum of plate and screen current. The red dot indicates that for a -26V gride and 365V "plate" the "plate current" is 46mA.

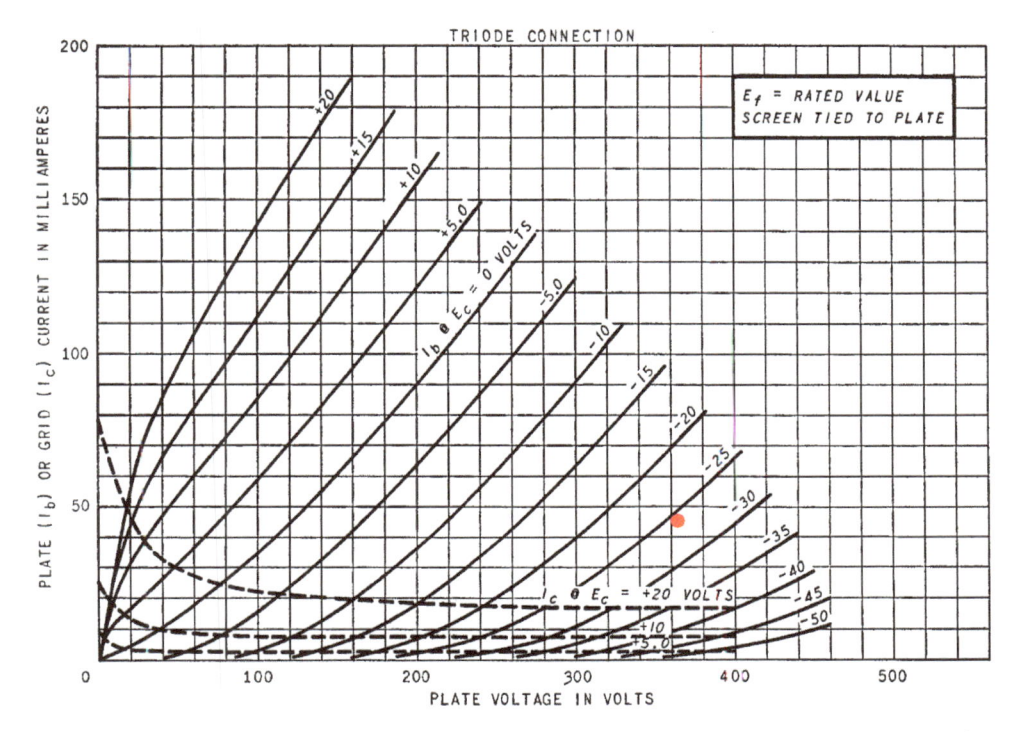

We determine later in this chapter that the total DC current load from the voltage amplifiers is 1.9mA and for the phase inverter it is 2.1mA.

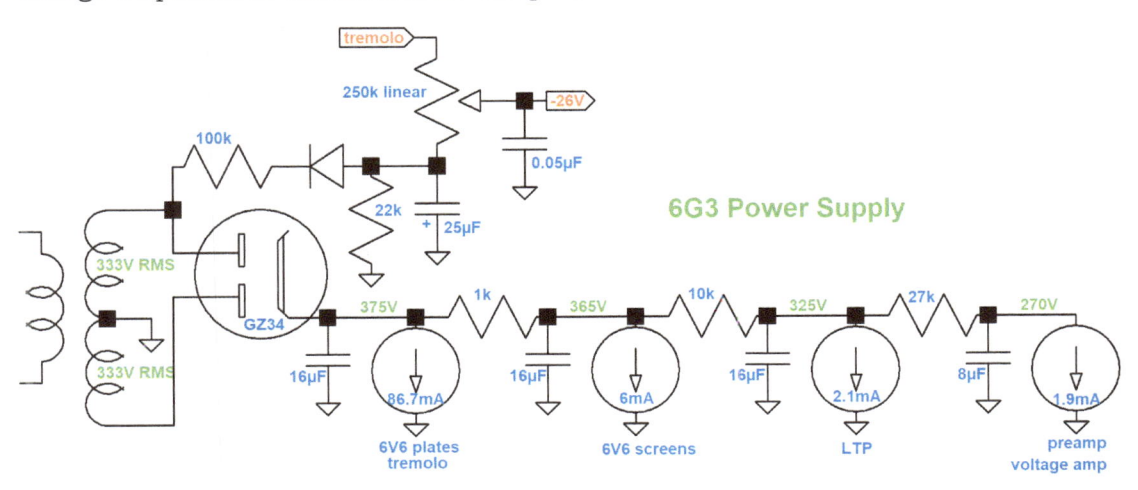

According to Ohm's Law, the total current through the 1kΩ ripple filter resistor is

$$\frac{375V - 365V}{1k\Omega} = 10mA$$

**ENHANCEMENT – THE BROWNFACE DELUXE 6G3 – 103**

This means the total screen current is somewhere in the neighborhood of $10mA - 4mA = 6mA$, which is 3mA per screen. Given Fender's usual plus or minus 20-percent accuracy for voltage measurements, this estimate is not particularly accurate, but good enough for evaluating power amp performance. Plate current for each 6V6 is the cathode current minus the screen current: $46mA - 3mA = 43mA$. The tremolo draws 700mA at 375V, as will be seen.

The 5E3's 1.5kΩ grid-stopper resistors are removed from the power amp and negative feedback is added between the output transformer secondary and the phase inverter.

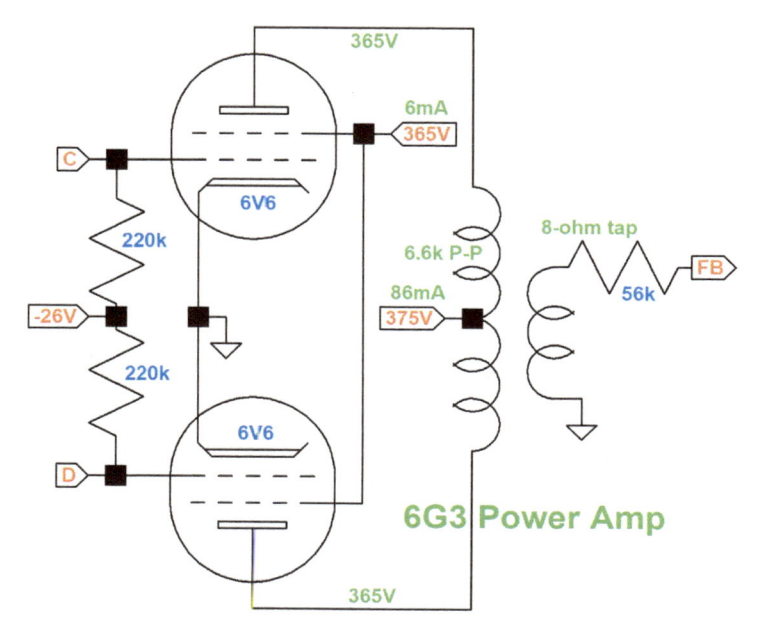

**6G3 Power Amp**

Input headroom matches the DC grid bias: 26V peak (+25.3dBV).

Fender measures a 10V drop across each phase of the output transformer primary. According to Ohm's Law, the DC winding resistance per phase is

$$\frac{10V}{43mA} = 233\Omega$$

For comparison, the DC winding resistance for a Hammond 1750H replacement transformer is 214Ω.

Idle plate dissipation is $(43mA)(365V) = 16W$ and idle screen dissipation is $(3mA)(365V) = 1.1W$. Maximum data sheet limits are 12W and 2W, respectively. The data sheet also specifies a maximum of 315VDC for the plate and screen. The 6G3 pushes its power tubes beyond their published

**104 – CHAPTER 4**

limits, but not nearly as much as blackface models, as we will see in the next chapter.

The DC plate voltage is 365V. If the voltage swings to 0V across an effective 1.65kΩ transformer primary impedance (6.6kΩ plate-to-plate), plate current swings to

$$\frac{365V}{1.65k\Omega} = 221mA$$

The endpoints for the AC load line, shown here in red, are therefore 0V, 221mA and 365V, 0mA.

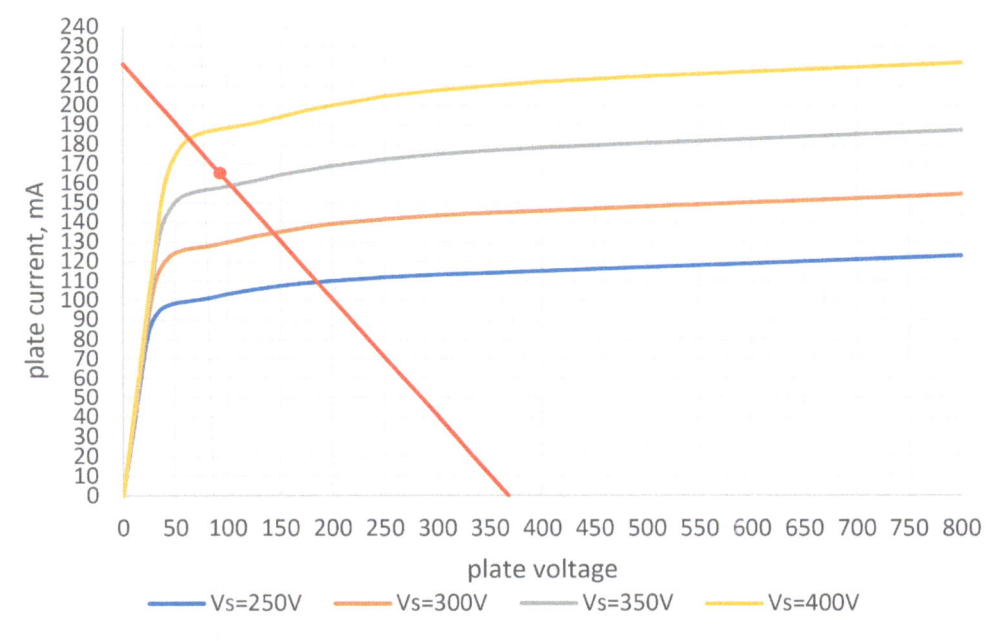

The load line intersects an imagined 0V curve for a 365V screen at 90V, 166mA, marked by the red dot, for an output power of approximately

$$\left(\frac{1}{2}\right)(365V - 90V)(166mA) = 23W$$

This estimate assumes the power supply voltages do not sag at full power. It represents 44-percent more power (1.6dB) compared to the 5E3. Increased power is one of Fender's main objectives and this design takes another step in that direction.

## Power Supply Modifications

For greater power, the power amp gets a higher screen voltage and fixed bias, effectively increasing the screen-to-cathode voltage by 66V. There is greater AC ripple at the screens, which fortunately are in a push-pull configuration for ripple cancellation. The long-tailed pair is also in a push-pull-like situation.

A common rule of thumb is to insert one ripple filter for every two inverting voltage amplifiers. As mentioned in Chapter 1, this keeps multiple stages from "motorboating." Presumably for this reason, Fender inserts another RC ripple filter between the first two stages and the phase inverter.

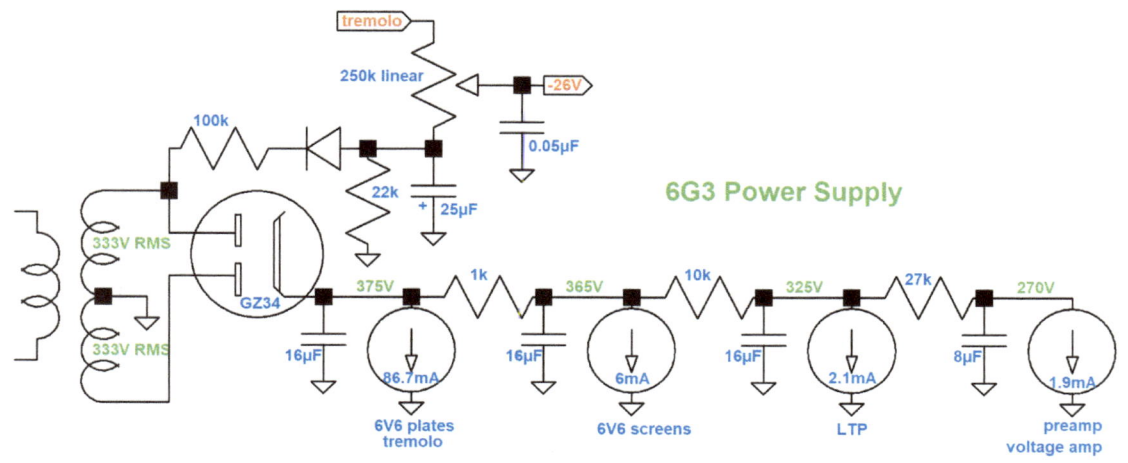

**6G3 Power Supply**

The 5Y3 rectifier used in the 5E3, with its high plate resistance, maximizes power supply squish and is a major contributor to the dynamics of a tweed Deluxe, particularly in the transition into and out of overdrive. The 6G3 is a recipient of Fender's "rectifier of the future,"[109] the GZ34. On the next page are the plate characteristics for a 5Y3 rectifier, as used in the Deluxe 5E3. The small inset in the lower right is a portion of the characteristics for a GZ34, as used in the 6G3.

As an illustrative example, the red dot shows that a 5Y3 requires 68V plate-to-cathode to conduct 200mA. The blue dot, on the other hand, shows that a GZ34 creates a voltage drop of only 14V in response the same load current. In the context of a dynamic musical performance, this difference is huge. For the 5Y3, changes in load current have a big impact on supply voltages. From a guitarist's perspective, a pick attack can start loud and crisp, swiftly back off in volume with increasing distortion as supply voltages sag, and then sustain for days as DC power slowly recovers.

> "If you take something like a tweed Deluxe and turn it all the way up and hit your low E string, that power supply just cannot keep up. You get that

---

[109] Tom Wheeler, **The Soul of Tone**, (Milwaukee: Hal Leonard, 2007), p. 86.

**106 – CHAPTER 4**

rectifier sag."[110] –Shane Nicholas, Fender Musical Instruments Corporation

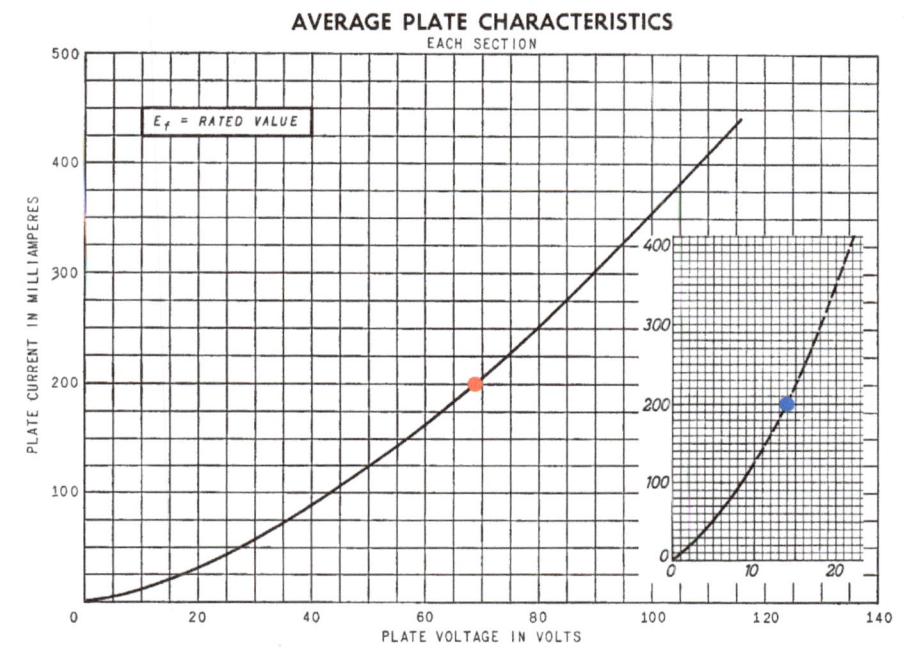

**AVERAGE PLATE CHARACTERISTICS**
EACH SECTION
$E_f$ = RATED VALUE

PLATE CURRENT IN MILLIAMPERES vs PLATE VOLTAGE IN VOLTS

From a purely technical perspective, on the other hand, the GZ34 has improved performance through greater rectification efficiency. Even its heater requirement is slightly more efficient: 1.9A at 5V versus 2A for the 5Y3.

> "Mr. Fender's quest to improve his amplifiers entailed much experimentation with rectifiers; he was well aware of evolving technologies in tube amplification outside the guitar business and strove to incorporate new developments to improve his own products; and finally, any rectifier arrangement that would enable the amp to deliver more clean volume was the way to go."[111]

For a DC load of 96.7mA at 375V, a 16µF capacitor input filter, and a ripple frequency of 120Hz, the ratio of the ripple voltage in RMS to DC voltage is[112]

$$\frac{\sqrt{2}(96.7mA)}{2\pi(120Hz)(375V)(16\mu F)} = 0.03$$

---

[110] Tom Wheeler, **The Soul of Tone**, (Milwaukee: Hal Leonard, 2007), p. 182.
[111] Tom Wheeler, p. 182.
[112] **Silicon Rectifier Handbook**, (Bloomington: Sarkes Tarzian, 1960), pp. 41, 42.

This means the ripple level is $(0.03)(375V) = 11.3V\ RMS$, which is 16V peak (+21.1dBV).

Ripple attenuation for the first RC filter, with a 1kΩ resistor and 16μF capacitor, is

$$\frac{1}{\sqrt{[2\pi(120Hz)(1k\Omega)(16\mu F)]^2 + 1}} = 0.0826\ (-21.7dB)$$

AC ripple at the 6V6 screens is therefore

$$21.1dBV - 21.7dB = -0.6dBV\ (1.3V\ peak)$$

Ripple attenuation for the second RC filter, with a 10kΩ resistor and 16μF capacitor, is

$$\frac{1}{\sqrt{[2\pi(120Hz)(10k\Omega)(16\mu F)]^2 + 1}} = 0.00829\ (-41.6dB)$$

AC ripple for the phase inverter is therefore

$$-0.6dBV - 41.6dBV = -42.2dBV\ (11mV\ peak)$$

Ripple attenuation for the third RC filter, with a 27kΩ resistor and an 8μF capacitor, is

$$\frac{1}{\sqrt{[2\pi(120Hz)(27k\Omega)(8\mu F)]^2 + 1}} = 0.00614\ (-44.2dB)$$

AC ripple for the preamps is

$$-42.2dBV - 44.2dBV = -86.4dBV\ (68\mu V\ peak)$$

## Introducing a DC Bias Supply

The grid bias supply[113] is a simple half-wave rectifier circuit that taps into one phase of the power transformer secondary, as shown on the next page. The diode's orientation and the voltage divider formed by the 100kΩ and 22kΩ resistors convert 471VAC peak (333V RMS) at the power transformer secondary to -26VDC. The 25μF capacitor acts as a ripple filter. (Its positive terminal is grounded, which is opposite the polarity of the high-voltage supply.)

The 250kΩ "intensity" control dials in more or less tremolo from the tremolo circuit, as described later in this chapter. The small capacitor at the output smooths out glitches from the tremolo circuit.

---

[113] Richard Kuehnel, **Guitar Amplifier Electronics: Basic Theory**, (Seattle: Amp Books, 2018), pp. 161-163.

## 6G3 Bias Supply

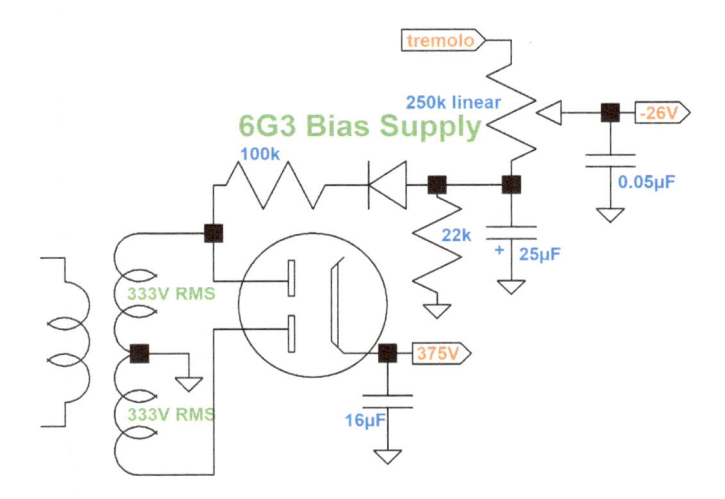

*[Schematic: 6G3 Bias Supply with tremolo input, 250k linear pot, -26V, 0.05µF, 100k, 22k, 25µF, 333V RMS windings, 375V, 16µF]*

## Introducing a Long-Tailed-Pair Phase Inverter

The 5E3 split-load phase inverter is replaced by a *long-tailed pair* (LTP), first seen in Fender's big amps of the late 1950s,[114] including the Bassman 5F6 and Twin 5F8.

Here is a simplified version of the 6G3 phase inverter.

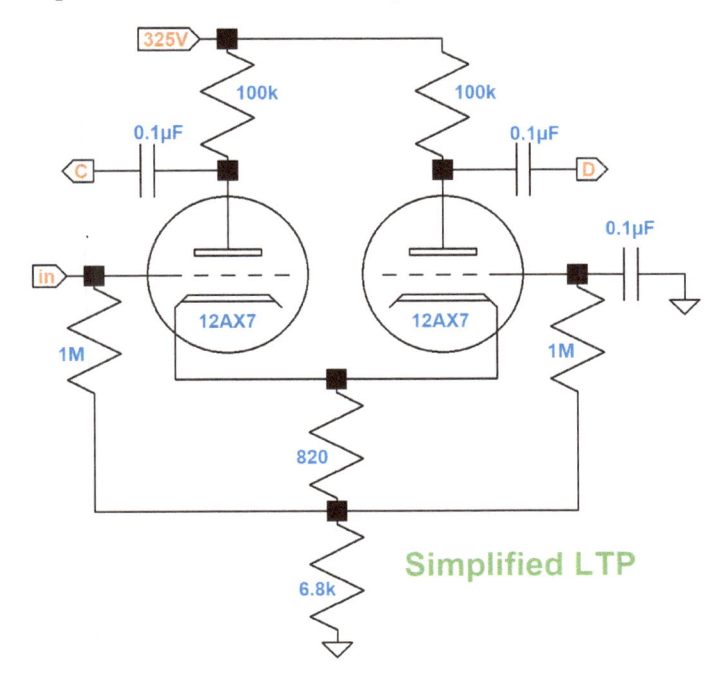

*[Schematic: Simplified LTP with 325V supply, two 100k plate resistors, two 12AX7 triodes, 0.1µF coupling caps to C and D, 1M grid resistors, 820Ω and 6.8k tail resistors, in input, 0.1µF]*

---

[114] John Teagle and John Sprung, **Fender Amps: The First Fifty Years**, (Milwaukee: Hal Leonard, 1995), p. 45.

When the input voltage at the left increases, it sets off simultaneous events that affect both triodes.

| Left Triode | Right Triode |
|---|---|
| Grid-to-cathode voltage increases. | |
| Plate current increases. | Total current through the 820Ω and 6.8kΩ resistors increases. |
| Voltage across the plate load resistor increases. | Voltage across the cathode resistor increases. |
| Output voltage at the plate decreases. | Grid-to-cathode voltage decreases. |
| **Input and output are of opposite phase.** | Plate current decreases. |
| | Voltage across the plate load resistor decreases. |
| | Voltage at the plate increases. |
| | **Output is in phase with the input.** |

A guitar signal enters from the left. A 0V audio "signal" enters from the right through a large coupling capacitor. The result is a pair of identical signals of opposite phase to drive the push-pull power amp.

If the plate load resistors have equal values, the inverted output at node C has more gain than the non-inverted output at node D. The 6.8kΩ *tail* could be made *longer* by increasing its value, thereby improving balance between the two outputs, but there is a tradeoff: increasing the size of the tail reduces maximum output voltage swing. If the tail is too long, the swing may be insufficient to drive the power amp to full power, particularly if the plate supply voltage is low.

Fender amps take a different approach to reduce imbalance: they reduce the value of the plate load resistor for the inverted phase to 82kΩ, as shown on the next page. This reduces inverted gain and increases non-inverted gain.

It is interesting to note that reducing the first plate load resistor value to 82kΩ is overcompensation. Back in the 1950s, a 91kΩ resistor with a 5-percent tolerance was uncommon. Nowadays a traditional 82kΩ value is usually retained to preserve the sonic legacy that Fender unleashed. The remaining imbalance drives the grid of one power tube to a greater amplitude than the other, creating 2nd-harmonic distortion in the power amp and a reverence for the tone of a classic Fender LTP.

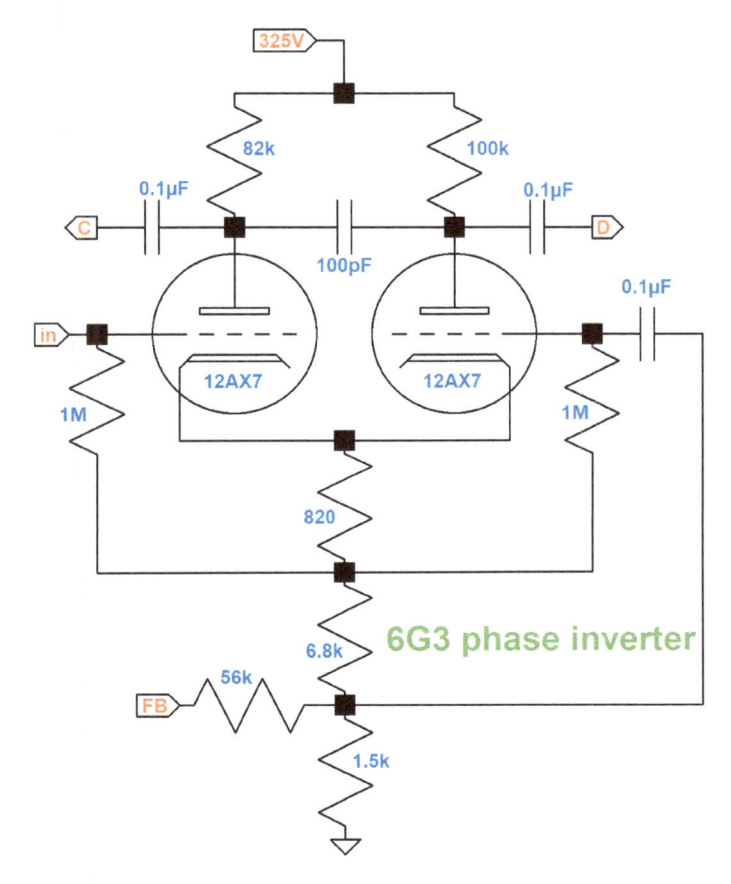

**6G3 phase inverter**

Both voltage gains are only about half the voltage gain of a triode amplifier stage. This is because grid-to-cathode voltage swing is only about half the input grid-to-ground voltage swing.

The 6G3 injects negative feedback from the output transformer secondary into the tail via a 56kΩ feedback resistor.

A 100pF capacitor is inserted between the plates to prevent radiofrequency oscillation. The noninverting grid is no longer connected to ground, as it was for our simplified version. It is instead connected to the feedback insertion point between the 6.8kΩ and 1.5kΩ resistors.

At DC the capacitors are open circuits. The resistors between the cathodes and ground carry the current of two triodes, so the equivalent resistances for one triode are doubled: 820Ω, 6.8kΩ, and 1.5kΩ for two triodes become 1.64kΩ, 13.6kΩ, and 3kΩ for one triode.

**ENHANCEMENT – THE BROWNFACE DELUXE 6G3 – 111**

The output transformer secondary has a very low DC resistance to ground, so the 56kΩ feedback resistor is in parallel with the 1.5kΩ resistor. The latter, however, is much smaller, so we can ignore the feedback resistor for DC. For the single-triode equivalent circuit, let's use 91kΩ for the plate load resistor, representing the average of the two resistor values.

If the plate current is 0mA, the plate-to-cathode voltage is equal to the plate supply voltage: 325V. According to Ohm's Law, if the plate-to-cathode voltage is 0V, the plate current is

$$\frac{325V}{91k\Omega + 1.64k\Omega + 13.6k\Omega + 3k\Omega} = 3mA$$

The endpoints of the DC load line are therefore 325V, 0mA and 0V, 3mA as shown here in red.

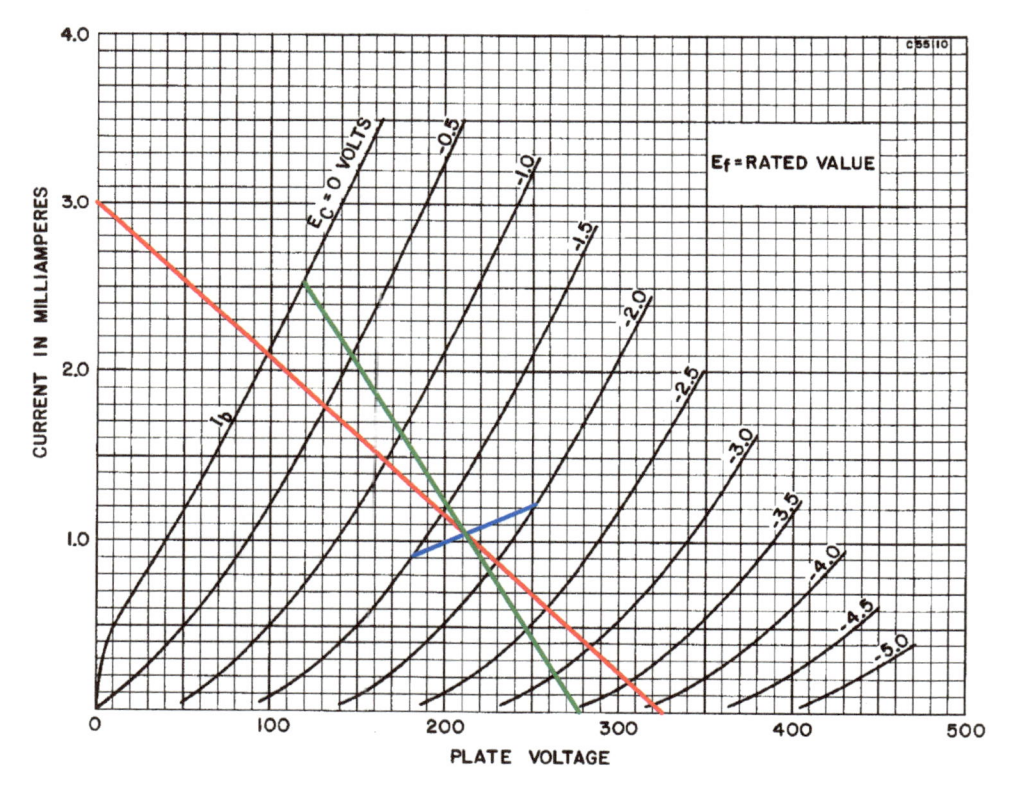

If the grid-to-cathode voltage is -1.5V or -2V, then according to Ohm's Law, the plate current is

$$\frac{1.5V}{1.64k\Omega} = 0.91mA$$

**112 – CHAPTER 4**

$$\frac{2V}{1.64k\Omega} = 1.22mA$$

These are the endpoints of the blue line segment. The intersection of the lines indicates that the DC operating point is defined by a grid-to-cathode voltage of -1.7V (Fender measures -2V.), a plate-to-cathode voltage of 211V (Fender measures an average of 208V.), and a plate current of 1.05mA.

DC plate current through the 82kΩ plate load resistor is slightly higher than 1.05mA and for the 100kΩ plate load resistor it is slightly less, but not enough to worry about. Based on Fender's measured voltages, the currents are 1.2mA and 1mA, respectively.

For audio signals the 91kΩ average plate load resistance is in parallel with a 220kΩ grid leak resistor for an effective AC plate load of 64kΩ. According to Ohm's Law, when the plate current swings from its DC value of 1.05mA to 0mA, the plate voltage swings by

$$(1.05mA)(64k\Omega) = 67V$$

The plate voltage therefore swings from its DC value of 211V to a value of $211V + 67V = 278V$. The AC load line therefore includes the DC operating point and the endpoint 278V, 0mA. The resulting AC load line is shown in green. Grid-to-cathode voltage swing is limited to 1.7V peak. The grid-to-ground swing is double, so input headroom is 3.4V peak (+7.6dBV). The plate voltage is limited to a swing of 67V peak. Input headroom for the power amp is only 26V peak, so phase inverter headroom is sufficient to drive the power amp to full power.

The loaded voltage gain is about half the voltage gain of a voltage amplifier with a 12AX7 amplification factor of 100, a 12AX7 plate resistance of 62.5kΩ, a 64kΩ AC plate load resistance, and a fully bypassed cathode resistor:

$$\left(\frac{1}{2}\right)\left(\frac{(100)(64k\Omega)}{64k\Omega + 62.5k\Omega}\right) = 25.3\ (28dB)$$

This assumes no feedback from the output transformer secondary and that the noninverting grid is connected to AC ground, as it was with our "simplified LTP" shown earlier. When so configured, the gain averages 28dB with about 0.1dB imbalance. Here is the response from the input to the inverted output (red) and non-inverted output (blue).

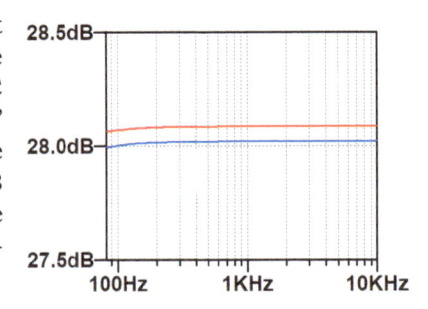

With Fender's tap into the tail for the noninverting grid, gain is reduced by less than a decibel.

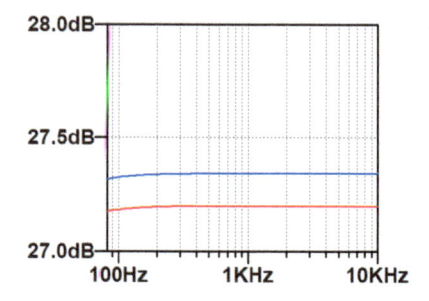

The voltage gain is 23 (27.2dB), a value that we will need for negative feedback calculations later in this chapter. The power amp needs +25.3dBV at the 6V6 grids to achieve full power. Without global negative feedback, the LTP input needs $25.3dBV - 27.2dB = -1.9dBV$.

## Introducing Negative Feedback

As the brownface amps are being designed, high fidelity is a popular concept, as is negative feedback for its implementation.

> "I am guessing that Leo preferred more negative feedback, and more speaker control, and less harmonic chaos – therefore 'higher fidelity.'"[115]
> –Shane Nicholas, Fender Product Development

(In reality, the relationship between feedback and "harmonic chaos" is not this simple, particularly with regard to intermodulation distortion when small amounts of feedback are applied. Nevertheless, Fender's philosophical intent is clear.)

Consider what happens when a voltage amplifier's cathode resistor is not bypassed by a large capacitor: a change in plate current through the cathode resistor RK causes a change in cathode voltage that is subtracted from the input voltage, creating a form of negative feedback called *cathode degeneration*.

In general, negative feedback takes a portion of the output signal and subtracts it from the input.

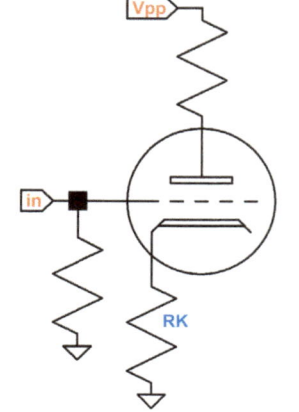

---

[115] Tom Wheeler, **The Soul of Tone**, (Milwaukee: Hal Leonard, 2007), p. 178.

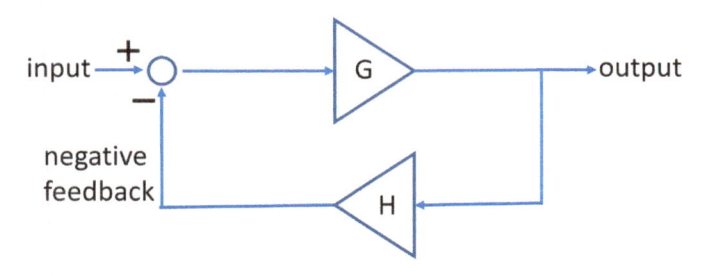

The *forward gain* G represents the gain from input to output that is present without negative feedback. The *feedback gain* H is from the output to the input. It is common to speak of *closed-loop gain*, which is the gain from input to output that includes the feedback loop. By contrast, forward gain is often called *open-loop* gain.

The feedback gain H can be frequency selective, which is the principle behind a *presence control*.[116] High frequencies are attenuated in the feedback loop, which boosts high-frequency closed-loop gain. For guitar players, the presence knob operates like a bright control. You can crank up the control to continue reining in bass using negative feedback, while allowing treble to run open and free.

Closed-loop gain is determined by a classic formula that is permanently embedded in the minds of electrical engineers, rocket scientists, and anyone else who deals with feedback and control systems:

$$G_{CL} = \frac{G}{1 + HG}$$

The formula shows that when negative feedback is added to a circuit, the gain is reduced by a factor of $1 + HG$. To determine closed-loop gain, we compute the individual gains that contribute to the forward gain $G$, then do the same for the feedback gain $H$, and then plunk these two values into the formula.

The driving stage, the power amp, the output transformer, and the feedback network form a system that can be broken down into the individual components that comprise $G$ and $H$.

---

[116] Richard Kuehnel, **Guitar Amplifier Electronics: Basic Theory**, (Seattle: Amp Books, 2018), pp. 146-147.

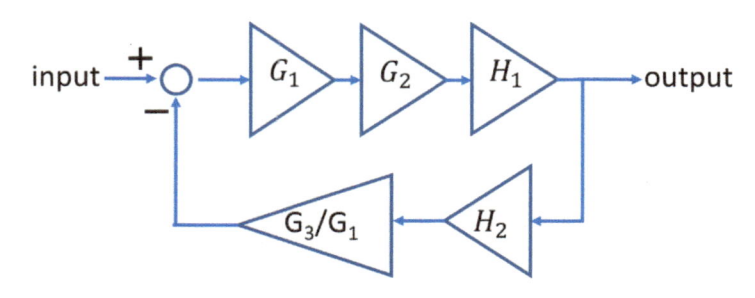

| | | |
|---|---|---|
| $G_1$ | Driving Stage Gain | LTP voltage gain $G_1$ includes the AC load of the power amp grid-leak resistors. |
| $G_2$ | Power Amp Gain | Power amp voltage gain $G_2$ is measured from the power tube grids to the transformer primary. |
| $H_1$ | Transformer Gain | $H_1$ represents voltage attenuation, because the transformer transforms the high-voltage, low-current environment of the primary circuit to the low-voltage, high-current environment of the speaker circuit. |
| $H_2$ | Feedback Gain | The voltage "gain" $H_2$ from the secondary to the LTP feedback input is the attenuation from a voltage divider formed by the feedback resistor and the shunt resistor connecting the LTP feedback input to ground. |
| $G_3$ | Driving Stage Feedback Input Gain | For an LTP phase inverter, the feedback voltage is fed to the grid of the triode opposite the signal input. This makes the phase inverter feedback voltage gain $G_3$ approximately equal to $G_1$. |

## How Negative Feedback Reduces Distortion

Power amp distortion occurs in the push-pull plate circuit, as shown here by the red arrow.

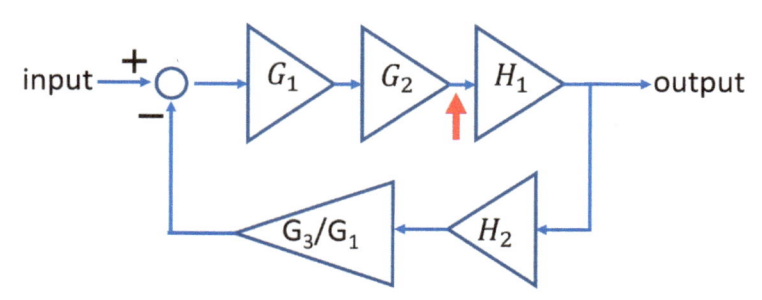

116 – CHAPTER 4

The loop for distortion is the same as for the guitar signal except for one key difference: the signal input relative to the signal output has changed. For the guitar signal input, the gains $G_1$ and $G_2$ are part of the forward path. For power amp distortion, they are part of the feedback path. From the distortion input to the output the forward gain is

$$G = H_1$$

From the output to the distortion input, the feedback gain is

$$H = H_2 \left(\frac{G_3}{G_1}\right) G_1 G_2 = H_2 G_3 G_2$$

Compared to the guitar signal, power tube distortion gets less forward-gain boost and more feedback gain attenuation.

## Feedback Calculations for the 6G3

The 6G3 feedback signal is taken from the same 8Ω tap that drives the speaker. It is attenuated by a voltage divider formed by a 56kΩ feedback resistor and the 1.5kΩ resistor at the end of the LTP tail. The attenuated signal is then fed to the grid of the triode driving the non-inverted output. This opposes the guitar signal at the opposite grid.

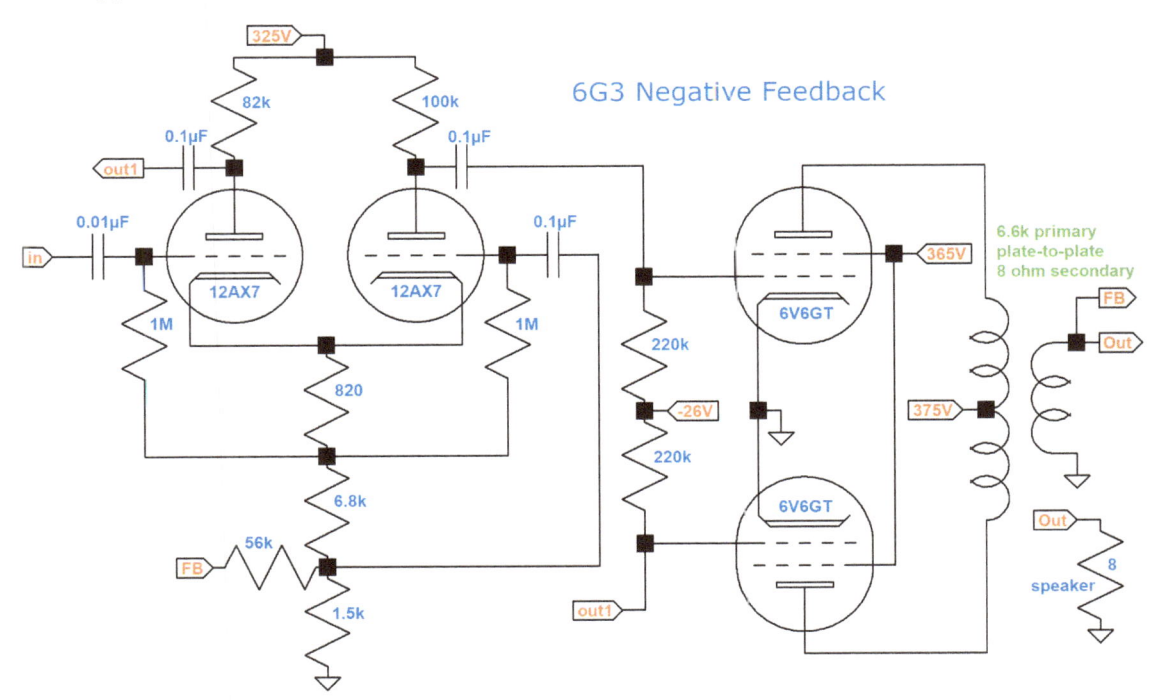

We determined in the last section that the inverted and non-inverted voltage gains for the LTP are approximately $G_1 = G_3 = 23$. At full power, the power amp produces about 23W RMS into an 8Ω load (as determined earlier without power supply voltage sag), so the RMS voltage measured at the speaker is

$$\sqrt{(23W)(8\Omega)} = 13.6V$$

The power tubes are biased at -26V. At full power, the RMS power amp input voltage is equal to the peak voltage divided by the square root of 2:

$$\frac{26V}{1.414} = 18.4V$$

The voltage gain for the power tubes and output transformer is the RMS voltage at the speaker divided by the RMS voltage at the power tube grids:

$$G_2 H_1 = \frac{13.6V}{18.4V} = 0.74$$

The phase inverter feedback voltage gain from the transformer secondary to the LTP feedback input is approximately the gain of the voltage divider formed by the 56kΩ feedback resistor and the 1.5kΩ resistor connecting it to ground:

$$H_2 = \frac{1.5k\Omega}{56k\Omega + 1.5k\Omega} = 0.026$$

Now that all the individual gains are known, we just crunch the numbers. The forward (open-loop) voltage gain is

$$G = G_1(G_2 H_1) = (23)(0.74) = 17 \ (24.6dB)$$

The feedback gain is

$$H = H_2 \frac{G_3}{G_1} = (0.026)\left(\frac{23}{23}\right) = 0.026$$

The closed-loop voltage gain, according to the famous feedback formula, is

$$G_{CL} = \frac{G}{1 + HG} = \frac{17}{1 + (0.026)(17)} = 11.8 \ (21.4dB)$$

The voltage gain with negative feedback is 69 percent of the voltage gain without feedback, i.e. 3.2dB less. This is only a modest reduction in gain, even for a guitar amplifier. We determined in the last section that input headroom is +7.6dBV without feedback. With feedback input headroom is

$$7.6dBV + 3.2dB = +10.8dBV \ (4.9V \ peak)$$

Without feedback the LTP needs -1.9dBV at its input to drive the power tubes to full power. With feedback, it requires

$$-1.9dBV + 3.2dB = +1.3dBV\ (1.6V\ peak)$$

This is substantially less than headroom, so the power amp reaches overdrive before the phase inverter.

The large 0.1µF coupling capacitors in front of the 6V6 grids create very little bass attenuation. Negative feedback flattens the bass response further. Here is the frequency response from the LTP input to the speaker with negative feedback (lower trace) and without negative feedback (upper trace).

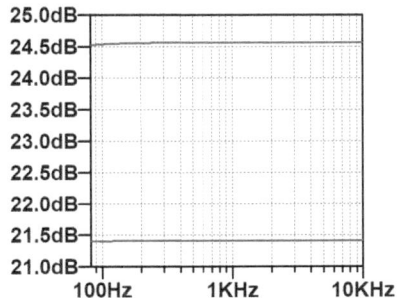

The upper trace shows just a tiny amount of attenuation at 82Hz, which almost disappears completely from the lower trace. Negative feedback flattens the response for all audio frequencies. It also reduces nonlinear distortion from the power amp and lowers the power amp's output impedance, making it more capable of handling the frequency-dependent nature of loudspeaker impedance, particularly for bass.

As mentioned in the last chapter, overdriving the power amp causes output clipping, which represents a lack of output response to a changing input signal voltage. The output provides the source voltage for negative feedback, so these effects reduce it. This creates more closed-loop gain, which drives the amp further into an overdriven state, producing even more clipping. The net result: as the amplifier is overdriven, the negative feedback circuit from the output transformer accelerates the transition to a more distorted state.

The brownface 6G3 has a modest amount of negative feedback that is comparable to early Marshalls, giving the amp a transition into overdrive that is less gnarly than a cranked-up tweed but not as "on the edge"[117] as blackface.

> "Most players rave about the sound of a cranked brownface Deluxe, which delivers a cool "mini-Marshall" approximation at manageable volumes."[118] –Dave Hunter

---

[117] Dave Hunter, **Amped**, (Minneapolis: Voyageur Press, 2012), p. 141.
[118] Dave Hunter, "Suhr Hombre: The Premier Guitar Review," **Premier Guitar**, March 24, 2020.

## An Additional Voltage Amplification Stage

The 5E3 has a second-stage voltage amplifier and a single-triode phase inverter with unity gain. The 6G3 uses a two-triode phase inverter with built-in gain but retains a modified second-stage voltage amplifier. In effect, the new design adds another voltage amplification stage. It has an unusual plate circuit, as shown here.

At DC the capacitors are open circuits and the grid is at ground potential via the upstream volume controls.

The DC load line extends from the plate supply voltage on the X axis to a Y axis value of

$$\frac{270V}{15k\Omega + 100k\Omega + 1.5k\Omega} = 2.32mA$$

as shown by the red line below.

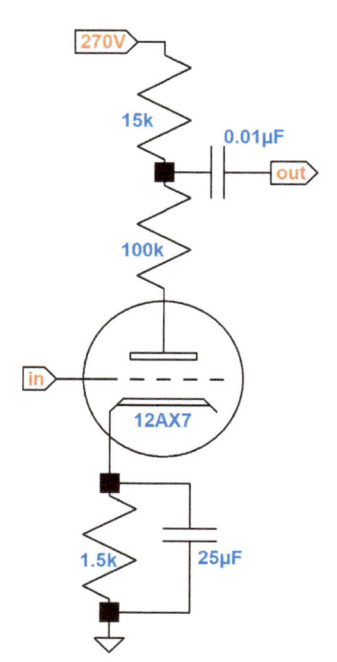

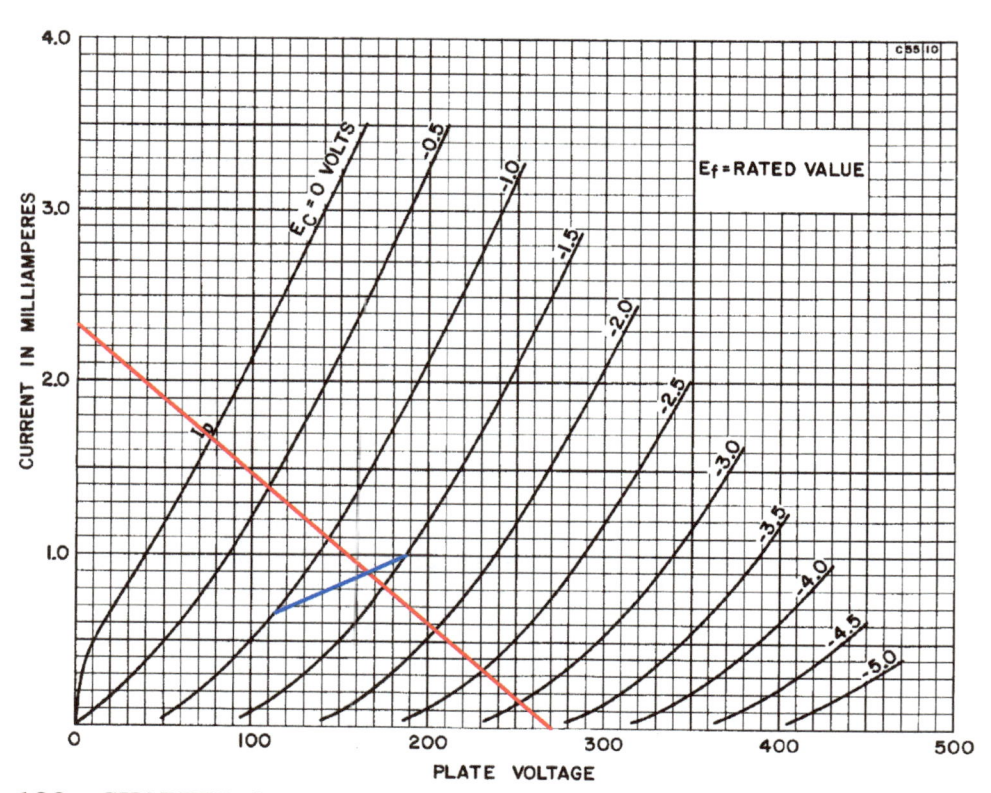

120 – CHAPTER 4

According to Ohm's Law, if the cathode voltages are 1V or 1.5V then the plate current values are

$$\frac{1V}{1.5k\Omega} = 0.67mA$$

$$\frac{1.5V}{1.5k\Omega} = 1mA$$

These values are the endpoints for the blue line segment. The point of intersection is the DC operating point: a grid-to-cathode voltage of -1.3V (Fender measures -1.4V), a plate-to-cathode voltage of 166V (Fender measures 164V), and a plate current of 0.9mA.

The LTP phase inverter represents a very light AC load, so the AC load line approximately coincides with the DC load line. The grid can swing by 1.3V before clipping, so input headroom is 1.3V peak (-0.7dBV).

For a 12AX7 amplification factor of 100, a 12AX7 plate resistance of 62.5kΩ, a plate load of 115kΩ, and a fully bypassed cathode resistor, the unloaded voltage gain measured at the plate is

$$\frac{(100)(115k\Omega)}{115k\Omega + 62.5k\Omega} = 64.8\ (36.2dB)$$

Because of the light AC load, this is approximately equal to the loaded gain.

A 15kΩ resistor is added to the top of the 100kΩ plate load resistor and their mutual connection is tapped for the output instead of the 12AX7 plate. This results in an intentional reduction in gain. For audio frequencies the 270V plate supply voltage is connected to ground via a large RC ripple filter capacitor, so for a guitar signal the output at the plate is attenuated by a voltage divider formed by a 100kΩ series resistor and a 15kΩ shunt resistor.

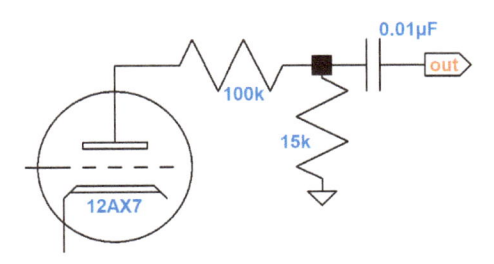

The resulting voltage divider has a "gain" of

$$\frac{15k\Omega}{100k\Omega + 15k\Omega} = 0.13\ (-17.7dB)$$

**ENHANCEMENT – THE BROWNFACE DELUXE 6G3 – 121**

Net gain from the voltage amplifier grid to the LTP grid is only $36.2dB - 17.7dB = 18.5dB$ and very flat.

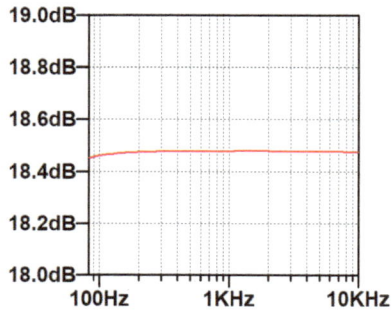

It is interesting to note that under these circumstances, the 25µF cathode bypass capacitor, which increases gain by eliminating cathode degeneration,[119] appears to be unnecessary – it can be eliminated by adjusting the voltage divider's resistor values for less attenuation. If the capacitor is removed, for example, and the plate load resistor and shunt resistor values are set to 75kΩ and 27kΩ, respectively, instead of 100kΩ and 15kΩ, the gain is the same: 18.5dB.

The LTP needs +1.3dBV at its input to achieve full power. At the second-stage grid, the required signal level is $1.3dBV - 18.5dB = -17.2dBV$ (195mV peak).

## More Aggressive Volume and Tone Controls

The 6G3 volume and tone control circuit is modified to be more aggressive, with increased insertion loss and much less interactivity. Each channel gets its own tone control in front of the second-stage voltage amplifier, as shown on the next page. The volume controls are top-fed and outputs are taken from the wipers, as is common in modern amplifiers. All controls have audio tapers.

The SPICE AC analysis simulation below shows the response from node A (normal channel, left plot) and node B (bright channel, right plot) to the grid of the voltage amplifier.

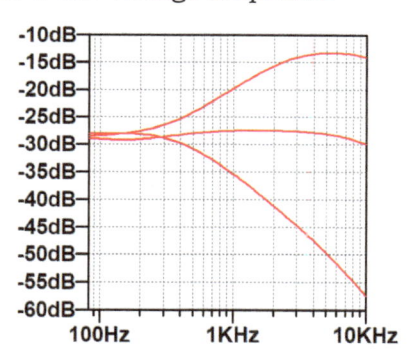

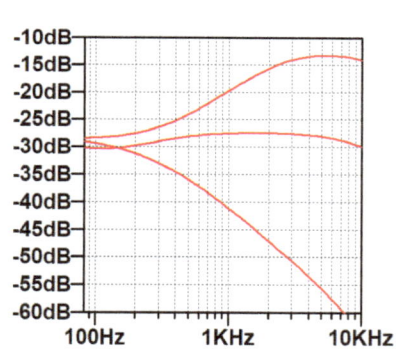

The active channel volume control is at 50-percent rotation (10-percent resistance) and the tone control is at minimum, 50-percent rotation (10-

---

[119] Richard Kuehnel, **Guitar Amplifier Electronics: Basic Theory**, (Seattle: Amp Books, 2018), p. 61.

percent resistance), and maximum. The plots assume the driving stage output impedance is 64kΩ, which we compute in the next section, and that the inactive channel volume control is at minimum.

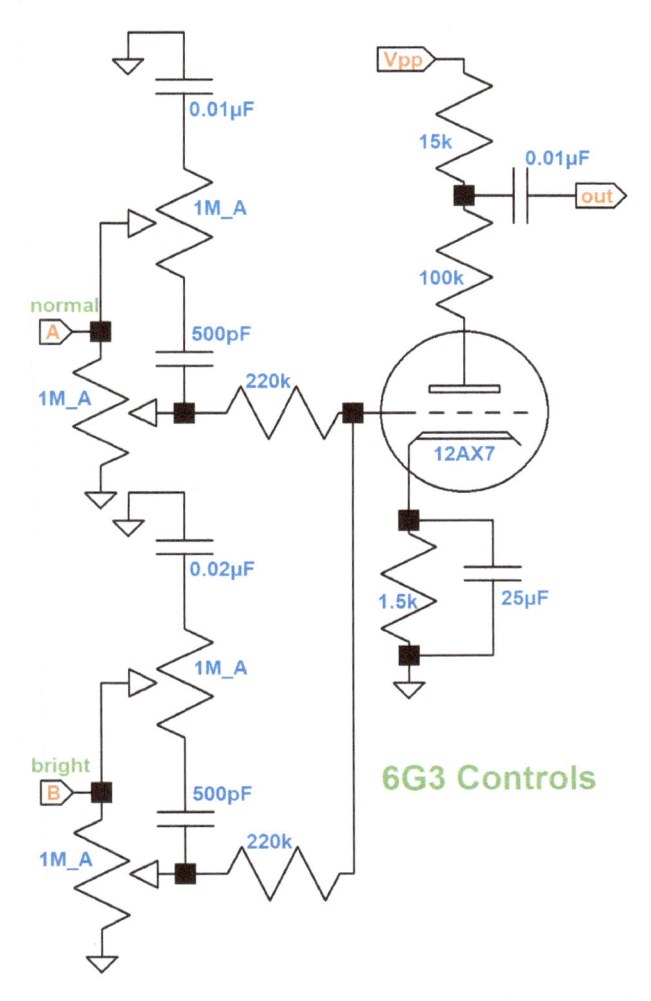

*6G3 Controls*

The subtle differences in response are due to the difference in tone control circuit capacitor values: 0.01µF for the normal channel and 0.02µF for the bright channel.

The tone control appears to provide treble boost, but the effect is an illusion. A volume control alone provides 20dB attenuation when set to 50-percent rotation. The tone control circuit adds more insertion loss and then allows treble to bleed through when the tone control is cranked up. For guitar amplifiers, tone shaping is almost always implemented with frequency-selective attenuation.

With the active channel's knobs at noon and the inactive channel's volume set to minimum, insertion loss is 27.5dB at 1kHz, assuming the driving circuit has a 64kΩ output impedance.

The second-stage grid needs a signal level of -17.2dBV to drive the power amp to full power. The unloaded signal level at the driving stage output needed for full power is

$$-17.2 dBV + 27.5 dB = +10.3 dBV \ (4.6V \ peak)$$

With the bright channel volume and tone controls at maximum, insertion loss is about 8.2dB and the required signal level is 19.3dB less:

$$-17.2 dBV + 8.2 dB = -9 dBV \ (502 mV \ peak)$$

## Increasing First-Stage Gain

For increased gain at the front end, a low-noise, low-hum 7025 with a 220kΩ plate load resistor replaces the 5E3's 12AY7 and 100kΩ resistor.

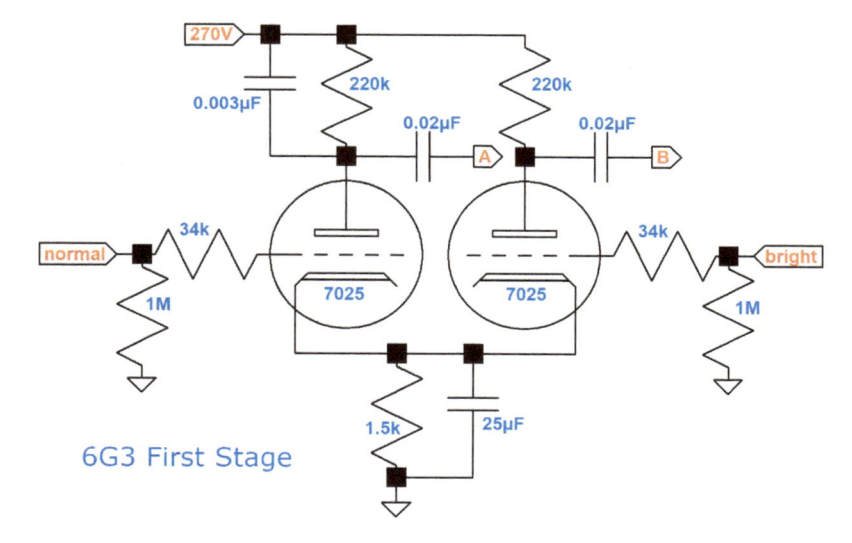

6G3 First Stage

"The popular, bright-sounding 7025 was intended as a low-noise, military/industrial equivalent to the 12AX7, and was introduced into the Fender line in the 1960's new Vibrasonic and Concert."[120] –Tom Wheeler

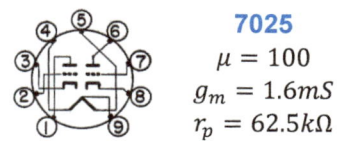

**7025**
$\mu = 100$
$g_m = 1.6 mS$
$r_p = 62.5 k\Omega$

---

[120] Tom Wheeler, **The Soul of Tone**, (Milwaukee: Hal Leonard, 2007), p. 85.

The tube swap and a doubling of the plate load resistor values boost amplification in front of the downstream attenuation caused by volume and tone controls. The inputs are now classified as "normal" and "bright," the former having a 0.003µF plate bypass capacitor[121] for treble cut.

For DC the capacitors are open circuits. The 1.5kΩ cathode resistor carries the cathode current of two triodes, so to create the same cathode voltage with only one triode its value needs to double: 3kΩ.

The red DC load line shown in red below extends from the 270V plate supply voltage on the X axis to a Y axis value of

$$\frac{270V}{220k\Omega + 3k\Omega} = 1.21 mA$$

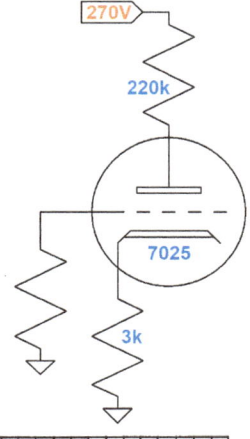

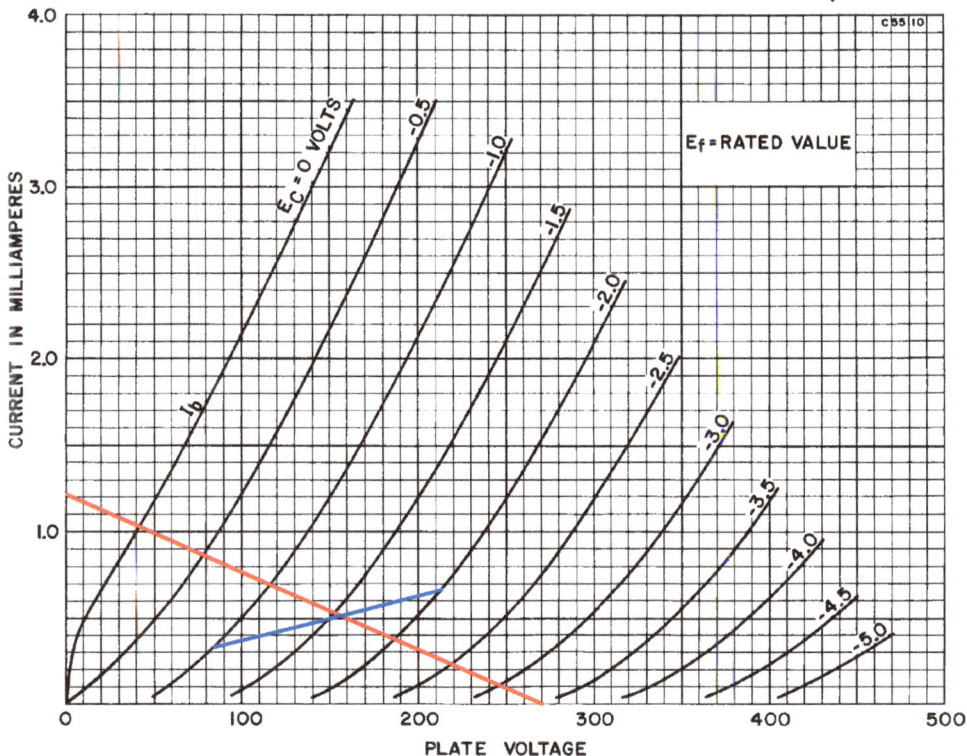

---

[121] Richard Kuehnel, **Guitar Amplifier Electronics: Basic Theory**, (Seattle: Amp Books, 2018), pp. 76-78.

According to Ohm's Law, if the cathode voltages are 1V, 1.5V, or 2V then the plate current values are

$$\frac{1V}{3k\Omega} = 0.33mA$$

$$\frac{1.5V}{3k\Omega} = 0.5mA$$

$$\frac{2V}{3k\Omega} = 0.67mA$$

These are the endpoints for the two line segments in blue. The point of intersection is the DC operating point: a grid-to-cathode voltage of -1.5V, a plate-to-cathode voltage of 156V (Fender measures 163V), and a plate current of 0.5mA.

The AC load varies considerably depending on frequency and knob position. The reactances of the normal channel's tone control capacitors for a 1kHz signal are

$$\frac{1}{2\pi(1kHz)(0.01\mu F)} = 16k\Omega$$

$$\frac{1}{2\pi(1kHz)(500pF)} = 318k\Omega$$

At 10-percent resistance, the significant resistances in the volume and tone control circuit are 100kΩ, so for the purposes of estimating a rough value for the AC load, let's consider the large capacitor to be a short circuit and the small capacitor to be open. This gives us the following circuit, comprised of the bottom 100kΩ for the volume control (RV2), the bottom 100kΩ for the tone control (RT2), and the 220kΩ mixing resistors (R1 and R2).

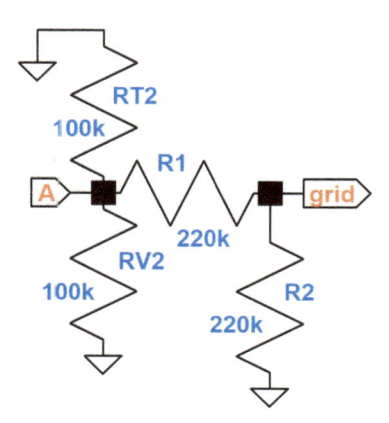

The grid is an open circuit when the second stage is not in overdrive, so the equivalent AC shunt resistance consists of three resistances in parallel: two 100kΩ resistances and a 440kΩ resistance:

$$\frac{1}{\frac{1}{100k\Omega} + \frac{1}{100k\Omega} + \frac{1}{440k\Omega}} = 45k\Omega$$

When the 220kΩ plate load resistor is in parallel with an AC load of 45kΩ the effective AC plate load is

$$\frac{1}{\frac{1}{220k\Omega} + \frac{1}{45k\Omega}} = 37k\Omega$$

According to Ohm's Law, when the plate current swings from its DC value of 0.5mA to 0mA, the plate voltage swings by

$$(0.5mA)(37k\Omega) = 19V$$

The plate voltage therefore swings from its DC value of 156V to a value of $156V + 19V = 175V$. The AC load line includes the DC operating point and the endpoint 175V, 0mA. The resulting AC load line is shown here in green.

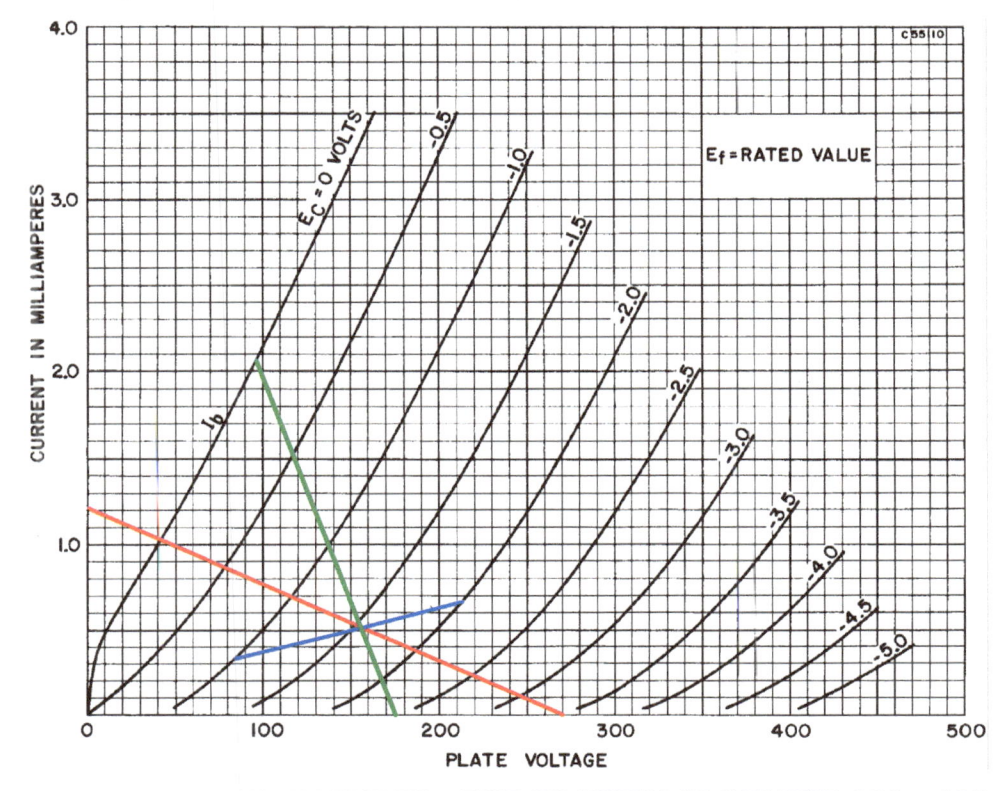

The grid voltage can swing by about 800mV peak before the stage breaks into severe distortion. Input headroom is therefore 800mV peak (-4.9dBV).

The DC plate current is only 0.5mA, so voltage gain is reduced. Based on a -1.5V grid and a 150V plate, a Sylvania 12AX7 data sheet indicates that the amplification factor is about 97 (blue dot) and the plate resistance is approximately 90kΩ (green dot).

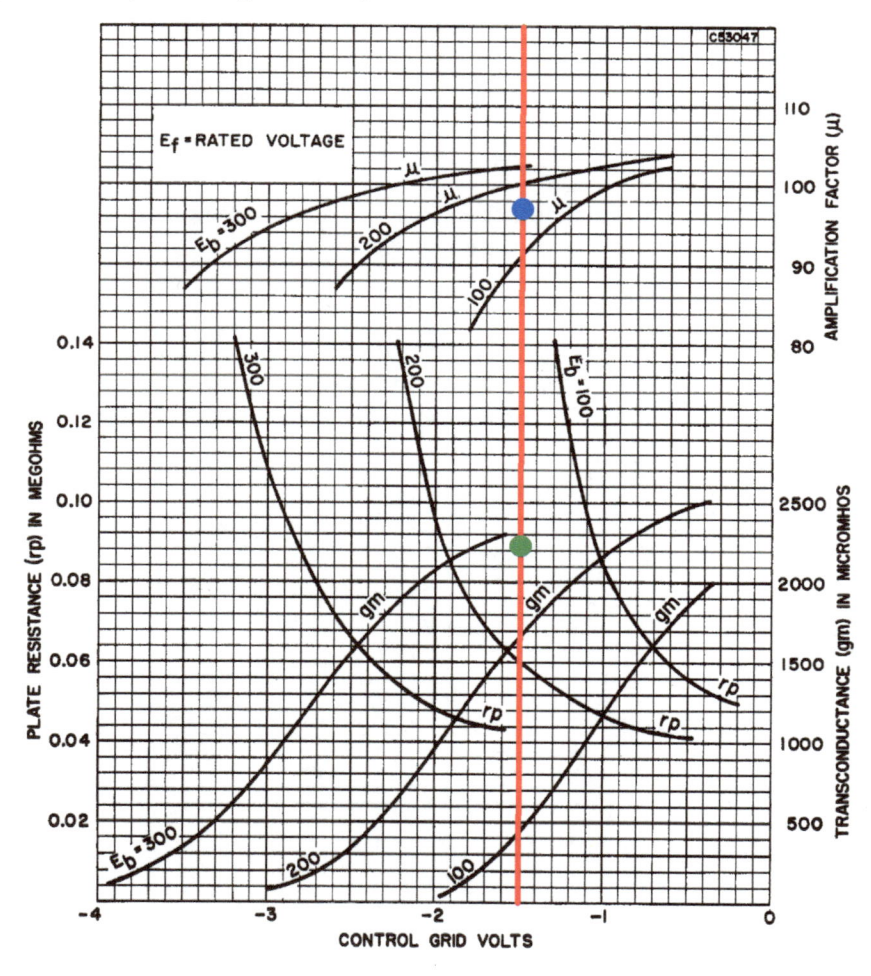

The unloaded voltage gain for a 220kΩ plate load resistor and a fully bypassed cathode resistor is

$$\frac{(97)(220k\Omega)}{220k\Omega + 90k\Omega} = 68.8 \ (36.8dB)$$

The output impedance, which we used in the last section, is

$$\frac{1}{\frac{1}{90k\Omega} + \frac{1}{220k\Omega}} = 64k\Omega$$

The unloaded signal level driving the volume and tone control circuit needs to be +10.3dBV for full power. The corresponding signal level at the guitar input jack is

$$10.3dBV - 36.8dB = -26.5dBV \; (67mV \; peak)$$

With the volume and tone controls cranked to maximum the required signal level for full power is

$$-9dBV - 36.8dB = -45.8dBV \; (7.3mV \; peak)$$

Both levels are within the capabilities of a single-coil pickup.

For the normal channel, the reactance of the 0.003µF plate bypass capacitor equals the 220kΩ plate load resistance at the break frequency of

$$\frac{1}{2\pi(64k\Omega)(0.003\mu F)} = 829Hz$$

This does not account for the AC load represented by the volume and tone control circuit. Here is the unloaded response from the guitar input jack to the plate. The upper trace is for the bright channel. The lower trace is for the normal channel.

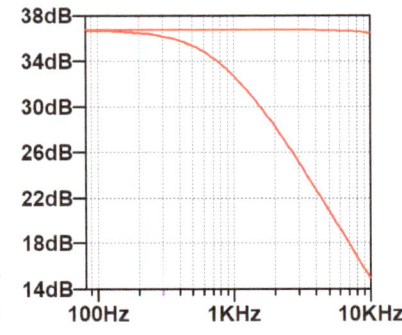

## Introducing Tremolo

The 6G3 tremolo circuit uses only one half of a 12AX7 dual triode, making the other triode available for the second-stage voltage amplifier described earlier. The circuit implements *grid-bias* or *bias-modulation* tremolo, a form of amplitude modulation that varies the loudness of the audio signal at a rate of a few hertz.

As a company, Fender has created confusion between *tremolo*, which modulates amplitude, and *vibrato*, which modulates pitch. What Fender calls "vibrato" is usually tremolo. There are a few brownface designs that, unlike the Deluxe tremolo circuit, "approach something akin to subtle pitch-modulating vibrato,"[122] but whether this is true vibrato is subject to debate.[123]

> "When you're talking about perception, I think the brown-era, multiple tube 'Vibrato' channel *did* in fact have a true vibrato. The phase shift caused a pitch shift, which is the definition of vibrato, so despite the conventional wisdom, I would call that circuit a phase-shift vibrato." –Matt Wilkens, Fender Principal Engineer
>
> "People sometimes call it a pitch shifting device. Technically, it's not, but the highs and lows do beat against each other, and you can perceive a little bit of that swirling effect." –Mark Baier
>
> "In fact, *no* Fender amp has *ever* had true pitch-bending vibrato, regardless of catalog hype to the contrary. It's fitting, in an ironic way, that this confusion of terminology should come from the company that calls the vibrato arm on its guitars 'tremolo'."[124] –John Teagle

A major contributor to the debate is perspective – physics notwithstanding, some of Fender's more complex brownface circuits sound like pitch bending to many ears.

The 6G3 tremolo circuit has a low-frequency oscillator (LFO) with a footswitch interface and a speed control.

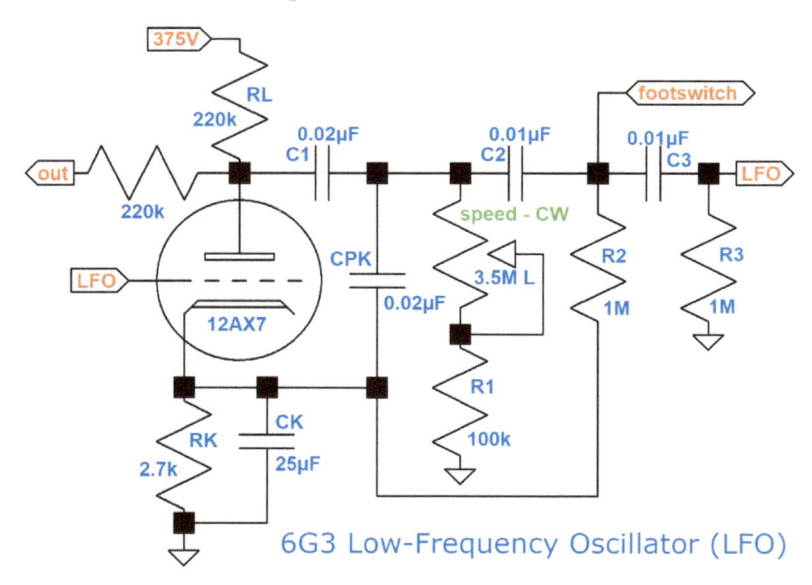

**6G3 Low-Frequency Oscillator (LFO)**

---

[122] Dave Hunter, **The Guitar Amp Handbook**, (Milwaukee: Backbeat Books, 2015), p. 55.
[123] Tom Wheeler, **The Soul of Tone**, (Milwaukee: Hal Leonard, 2007), p. 273.
[124] John Teagle and John Sprung, **Fender Amps**, (Milwaukee: Hal Leonard, 1995), p. 28.

The signal it generates, which has a frequency of a few hertz, is used to modulate the amp's volume, creating a throbbing effect synced to the LFO frequency.

At DC the capacitors are open circuits and the triode circuit looks like a typical voltage amplification stage with the grid at ground potential, a 220kΩ plate load resistor, a 2.7kΩ cathode resistor, and a 375V plate supply.

Using a traditional DC load line and grid line segments we can determine the DC operating point: a grid-to-cathode voltage of -2V (matching Fender's measurement), a plate-to-cathode voltage of 212V (Fender measures 203V), and a plate current of 0.73mA.

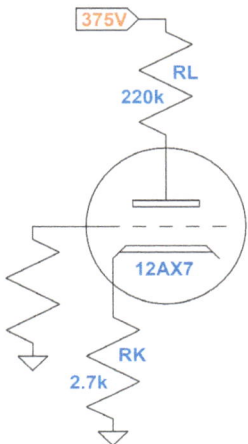

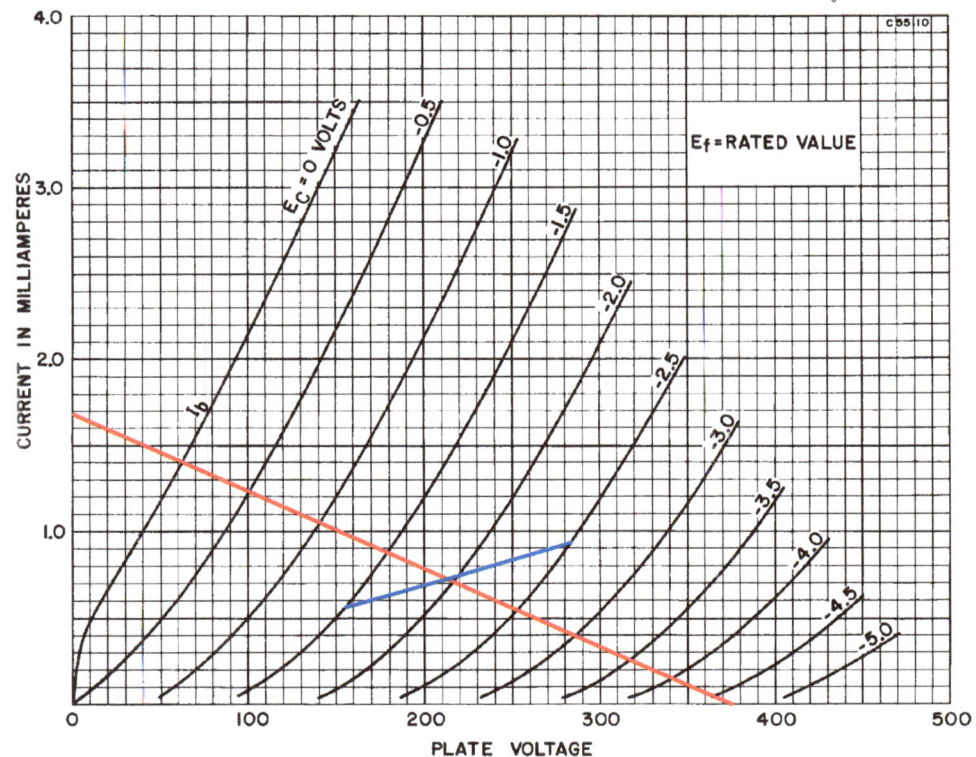

The unloaded voltage gain for a 12AX7 amplification factor of 100, a 12AX7 plate resistance of 62.5kΩ, a 220kΩ plate load resistor, and a fully bypassed cathode resistor is

**ENHANCEMENT – THE BROWNFACE DELUXE 6G3 – 131**

$$\frac{(100)(220k\Omega)}{220k\Omega + 62.5k\Omega} = 77.9 \ (37.8dB)$$

The voltage amplifier's AC load consists of a network of resistors and capacitors whose output (node LFO) is fed back to the grid for re-amplification.

Let's look at capacitor C3 and resistor R3 in isolation. Together they create an RC high-pass filter.[125] At low frequencies it introduces lots of attenuation (solid trace below) and a phase shift (dotted trace, scale on the right) of 90 degrees. At high frequencies the circuit has no attenuation and no phase delay, i.e. a phase shift of 0 degrees.

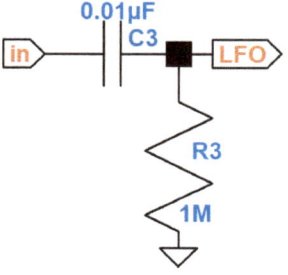

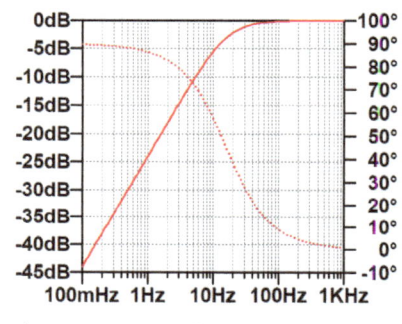

At 9Hz the attenuation is 6.2dB and the phase shift is 60 degrees. This is interesting, because it suggests that if we use three of these high-pass filters in series (and if they do not load each other down, which they do) then together they create 18.6dB attenuation and a phase shift of 180 degrees. The filters are not independent, so the actual frequency at which the output is 180 degrees out of phase with the input is lower. (SPICE simulation indicates it is 7.4Hz for the three filters in series.)

The amplifier that drives the three filters has a gain of well over 30dB, so the amplifier and filters work together to create substantial gain. Moreover, the amplifier is inverting – its output is 180 degrees out of phase with its input. From grid to plate the signal is shifted 180 degrees by the amplifier and then another 180 degrees by the filters for a net shift of 360 degrees. The filter output is therefore in phase with the amplifier's input and has a greater amplitude. This creates positive feedback to support oscillation.

The 3.5MΩ speed control varies the resistance for one of the RC filters, which varies the resonant frequency, i.e. the frequency at which the LFO output is in phase with the input. The 6G3 circuit also adds capacitor CPK to smooth out oscillator glitches.

To stop tremolo oscillation, the footswitch shorts the signal at the top of resistor R2 to ground. The bottom of R2 is connected to the cathode instead of ground. Since the cathode is fully bypassed by cathode resistor CK, even at tremolo frequencies, R2 is still connected to AC ground, so

---

[125] Richard Kuehnel, **Guitar Amplifier Electronics: Basic Theory**, (Seattle: Amp Books, 2018), p. 21.

oscillation performance is the same. The cathode voltage is 2VDC, however, so when the footswitch opens there is an initial AC jolt to get LFO oscillation started.

A 0.1µF capacitor connects the output to a 250kΩ intensity control.

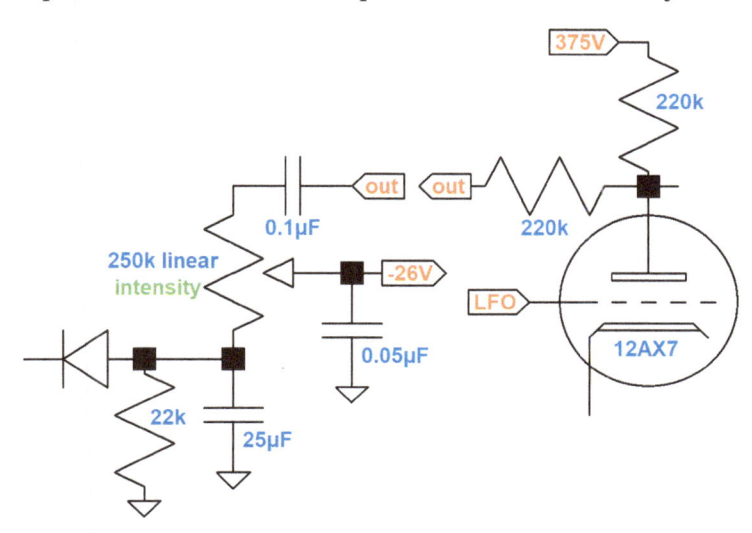

The capacitor blocks DC, so the intensity control sits between a pure AC signal from the LFO and a pure DC voltage from the DC grid bias supply. The bias for the 6V6 power tubes is therefore a combination of the two. When the control is fully clockwise, the -26VDC bias is AC modulated to its maximum extent. Even when cranked, however, negative swings of the bias voltage are insufficient to force the amp into Class B operation or beyond. This creates a softer, more nuanced character compared to an opto-cell circuit like the one in the Deluxe AA763.

## System Profile

Deluxe 6G3 system sensitivity is -26.5dBV (67mV peak). With both controls cranked to maximum, sensitivity is -45.8dBV (7.3mV peak).

The schematic on the next page documents what we now know about the 6G3 when the bright channel is active and the amp is at full power. On the page after that is a system profile for the bright channel.

The blue trace shows the signal level at the input to each stage at full power with the volume and tone controls at 50-percent rotation. The orange trace is with the controls at maximum. The yellow traces show input headroom for the first- and second-stage preamps, the phase inverter, and power amp. The gray trace shows power supply AC ripple for the plates, screens, phase inverter, and preamps. All traces are in dBV.

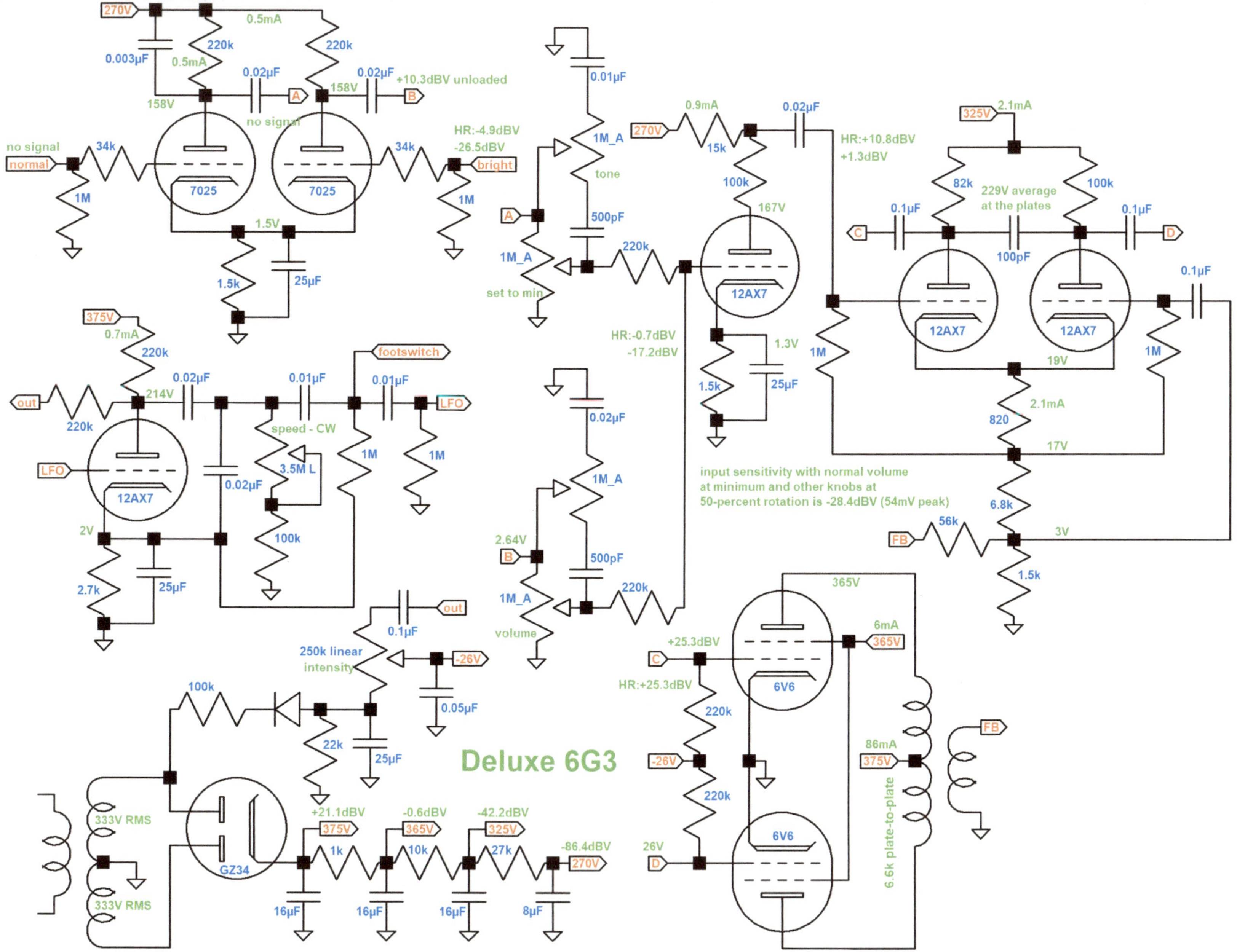
Deluxe 6G3
270V
0.5mA
220k
0.003µF
0.02µF
+10.3dBV unloaded
158V
no signal
normal
34k
1M
7025
1.5V
1.5k
25µF
HR:-4.9dBV
-26.5dBV
bright
375V
0.7mA
out
214V
0.01µF
footswitch
LFO
speed - CW
3.5M L
100k
12AX7
2V
2.7k
250k linear
intensity
0.1µF
-26V
0.05µF
22k
333V RMS
GZ34
+21.1dBV
-0.6dBV
-42.2dBV
-86.4dBV
375V
365V
325V
1k
10k
27k
16µF
8µF
1M_A
tone
500pF
set to min
A
B
2.64V
volume
220k
15k
0.9mA
167V
HR:-0.7dBV
-17.2dBV
1.3V
HR:+10.8dBV
+1.3dBV
2.1mA
82k
229V average
at the plates
100pF
C
D
19V
820
17V
6.8k
56k
FB
3V
input sensitivity with normal volume
at minimum and other knobs at
50-percent rotation is -28.4dBV (54mV peak)
+25.3dBV
HR:+25.3dBV
6V6
6mA
86mA
26V
6.6k plate-to-plate

## 6G3 System Profile

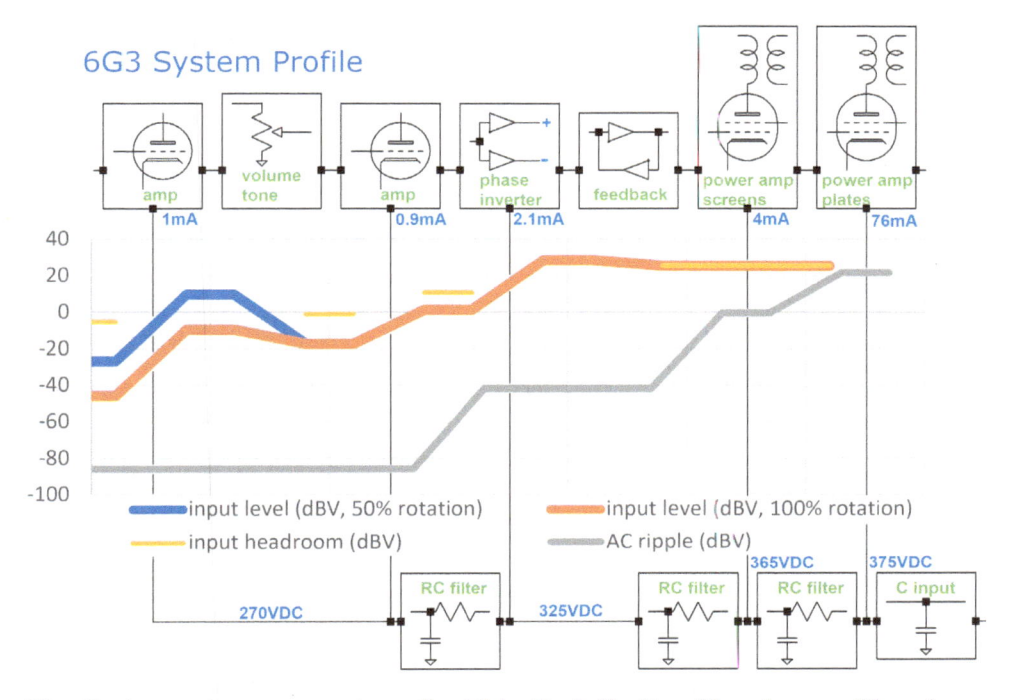

The first two stages operate well within their limits of headroom. The phase inverter has enough excess headroom to ensure that the power amp is well into overdrive before it begins to break up.

To drive the amp to full power, the 6G3 provides only slightly more gain than the 5E3. With the 6G3 controls at noon, for example, input sensitivity is 0.8dB lower than for the 5E3. With the knobs cranked, it takes only 3.2dB less guitar to drive the amp to full power compared to the 5E3, despite numerous changes to the signal chain. From a system design perspective this is an interesting observation that can provoke some speculation on Fender's design process.

The second stage has the potential to easily provide at least 35dB gain. Fender adjusts its built-in attenuator, however, to dial the gain down to only 18.5dB. We can surmise that Leo Fender implemented the power amp's more negative grid bias, added negative feedback, swapped out the phase inverter, implemented more aggressive controls, replaced the mild-mannered 12AY7 front end with an aggressive 7025/12AX7 preamp, and then, as a final step in his system design process, tweaked second-stage gain to provide approximately the same input sensitivity as the 5E3.

## System Voicing

With volume and tone controls at 50-percent rotation (10-percent resistance) for the active channel and the inactive volume at minimum, here is the response from the guitar input jack to the second 6V6 grid normalized to the gain at 820Hz.

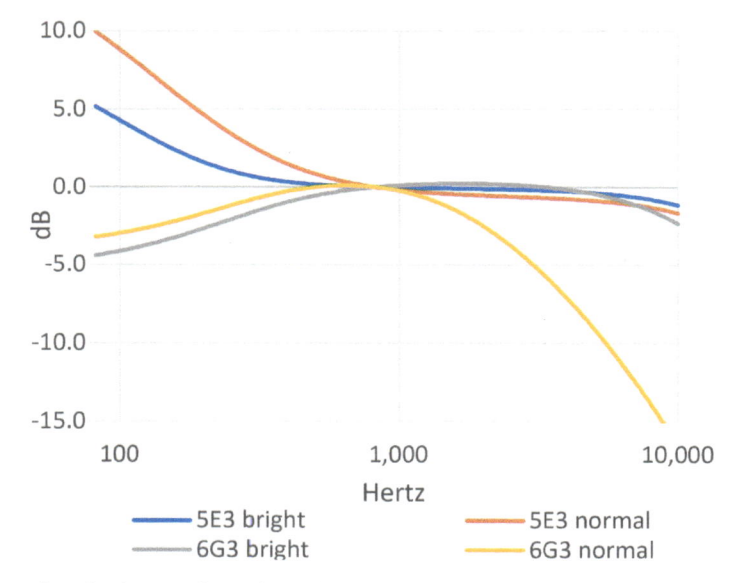

*Chart: dB vs Hertz (100 to 10,000), showing four curves: 5E3 bright, 5E3 normal, 6G3 bright, 6G3 normal.*

In terms of relative gain, the 5E3 varies the amount of bass boost to differentiate the normal and bright channels. The 6G3, on the other hand, voices the amp by controlling treble cut. Increased bass attenuation for the 6G3 is due to smaller coupling capacitor values.

For models up to and including the 5E3, Fender uses relatively large 0.05µF or 0.1µF coupling capacitors for the first stage. This shifts bass cutoff to lower frequencies. The 6G3 reduces these capacitor values to a more modern 0.02µF, which reduces the effective bass boost relative to treble. We can speculate that the reduction in bass emphasis may be influenced by musicians moving away from string bass towards electric bass, which can cover the first octave of a guitar, thereby relieving the guitarist from needing to fill that frequency range.[126]

## Full Power Performance

Screen-to-cathode voltage is considerably higher than in the 5E3. Here is a SPICE transient simulation[127] of one cycle of a 1kHz sine wave at full power.

---

[126] Personal correspondence with Paul Reid, February 2021.
[127] Richard Kuehnel, **Guitar Amplifier Electronics: Circuit Simulation**, (Seattle: Amp Books, 2019).

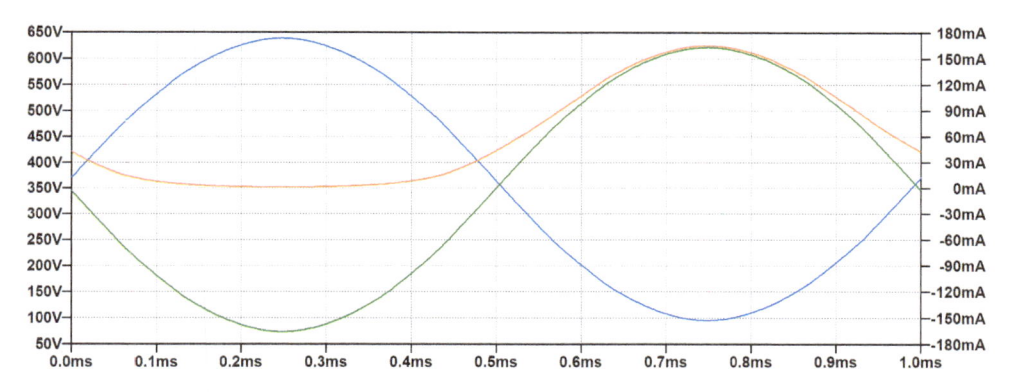

The traces show upper 6V6 plate voltage (blue), upper 6V6 plate current (orange), and net plate current (upper plate current minus lower plate current, green). Plate current and screen current average 64mA and 10mA, respectively, for each tube.

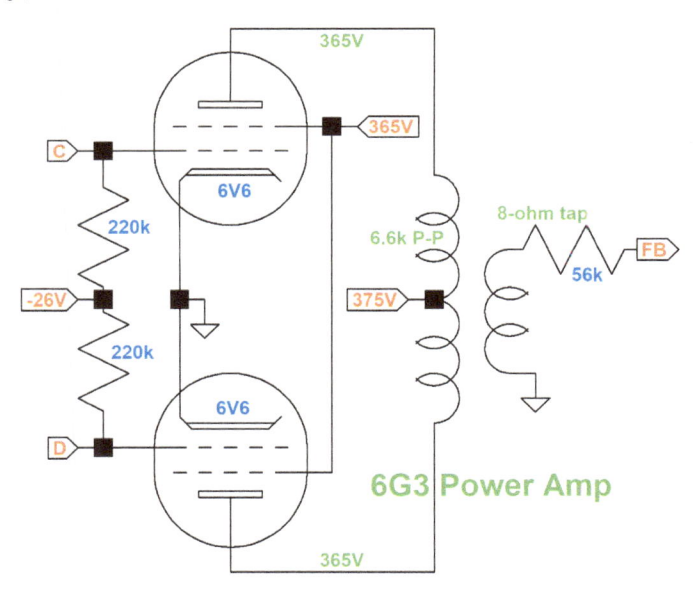

On the next page are the same performance curves we plotted for the 5E3, this time for a 365V screen. Upper tube plate current (yellow curve) swings down to only 2mA, almost to cutoff, so the 6G3 operates close to Class AB[128] but is still technically Class A, at least until supply voltages sag. The intersection of the yellow curve and the orange curve for a -26V grid represents the DC operating point: a plate voltage of 365V and a plate current of 43mA.

---

[128] Richard Kuehnel, **Guitar Amplifier Electronics: Basic Theory**, (Seattle: Amp Books, 2018), pp. 114-117.

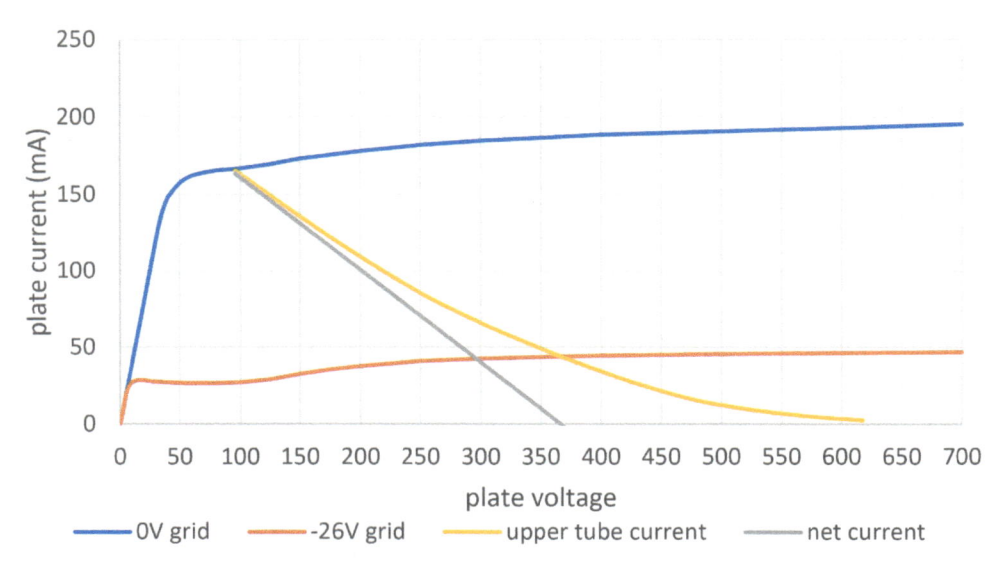

[Chart: plate current (mA) vs plate voltage, with curves for 0V grid, -26V grid, upper tube current, and net current]

The gray curve, representing the net plate current in the output transformer (upper tube plate current minus lower tube plate current) is 0mA when the plate voltage swings through its DC value of 365V. When the upper tube's grid swings to 0V, the plate voltage swings to 96V and the upper tube plate current swings to 165mA. The lower tube's grid swings to -52V for a plate current of 2mA. Net plate current is 163mA.

A plate voltage swing of $365V - 96V = 269V$ and a net plate current swing of 163mA represents an impedance of

$$\frac{269V}{163mA} = 1.65k\Omega$$

This is one fourth the plate-to-plate primary impedance of 6.6kΩ. Full power is approximately

$$\left(\frac{1}{2}\right)(269V)(163mA) = 21.9W$$

This is before power supply voltage sag. It also assumes an ideal output transformer and an 8Ω purely resistive speaker. The 6G3's higher screen supply voltage and fixed bias combine to create 1.4dB more power than the 5E3.

## Bias Excursion and Power Supply Sag

In overdrive, grid current causes the 0.1μF coupling capacitors to charge over the course of a few milliseconds. This makes the effective 6V6 DC grid bias more negative, creating bias excursion that pushes the tubes toward Class B operation.

Here is the power supply under idle conditions.

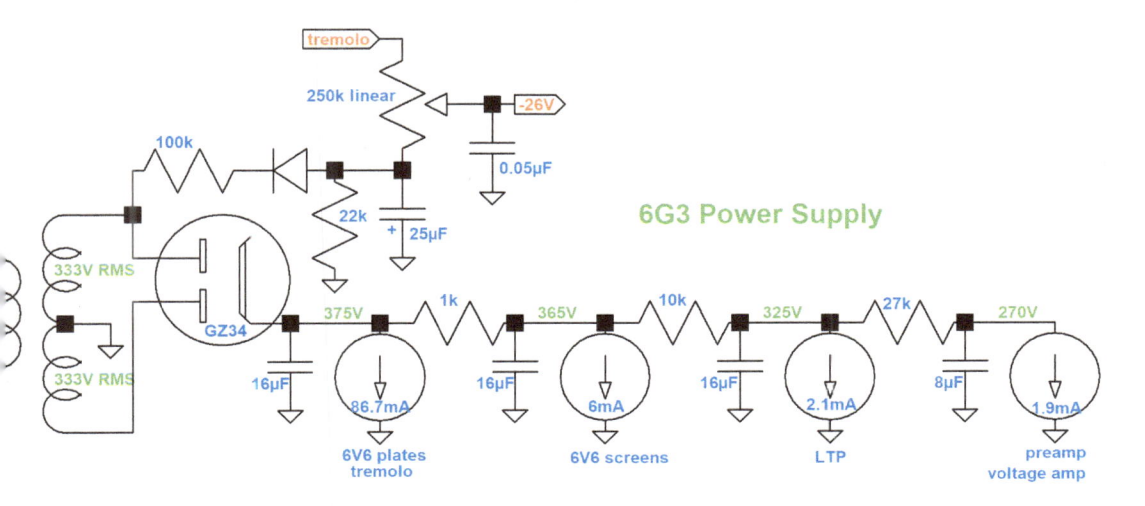

**6G3 Power Supply**

Downstream of the GZ34 there is a total load current of 97mA. The power supply DC output voltage is 375V, as marked by the red dot on the operation characteristics of a GZ34 shown on the next page. The curves are for a 60µF capacitor input. The Deluxe has only 16µF, which lowers the output voltage compared to what is depicted by the curves for a typical power transformer.

At full power, average 6V6 plate current for two tubes is 128mA. Average screen current for two tubes is 20mA. The total power supply load current is 159mA, so the plate supply sags by about 29V, as shown by the blue dot. This assumes, however, that average plate current and screen current hold steady over the course of the sag. In reality, full power load current decreases in response to lower supply voltages, so our dots represent boundaries for the power supply's operating range during a real musical performance. Overdrive creates even more sag.

140 – CHAPTER 4

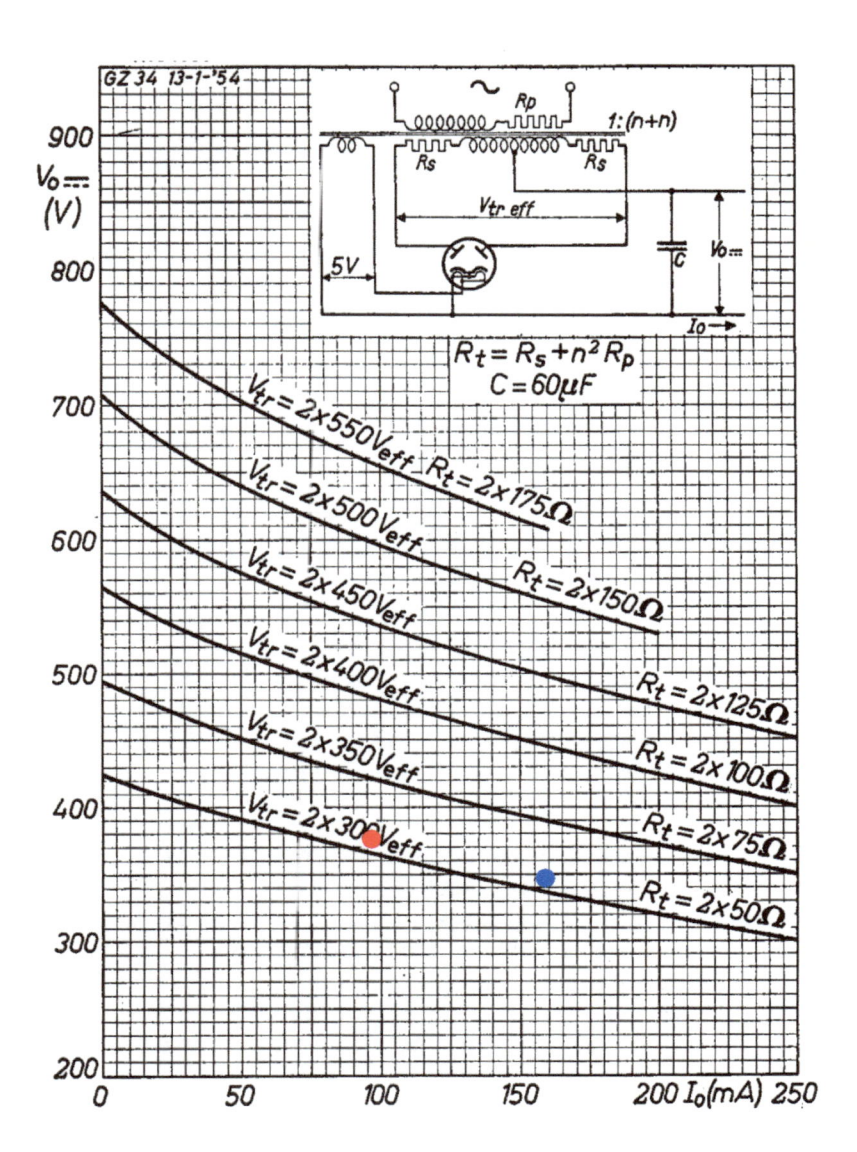

# Chapter 5: Perfection – The Blackface Deluxe Reverb

"Desert island amp – all you need right there..."[129]

—Joe Bonamassa

## AA763 System Design Concept

*Trickle-down economics* is a theory claiming that financial benefits for the wealthy eventually trickle down to the middle class. The validity of the theory is debatable.[130] It can be said, however, that many of the features introduced into Fender's high-end models eventually *trickle down* to the *middle class* of the company's product line. A full tone stack, for example, requires an investment in additional parts, both for the stack itself and for mitigating its increased insertion loss. An expensive stack is first featured in the Bassman 5F6 and Twin 5F8. Next in line are high-end brownface models. It is not until blackface that the upgrade trickles down to the Deluxe, giving it bass and treble controls just like its wealthier brethren.

> Fender accurately termed its separate controls for bass and treble "the latest in electronic advances."[131] –Tom Wheeler

Substituting a choke for a resistor in the first ripple filter is also an expensive upgrade. The Bassman 5D6-A is perhaps the first push-pull amp to get a choke. Next in line are high-end narrow panels: the Twin, Bandmaster, Pro, and Super. Brownface recipients include the Concert, Showman, Tremolux, Vibrasonic, Vibroverb, and Vibrolux. It is not until blackface that a choke finally trickles down to the Deluxe.

No-cost upgrades, by contrast, circumvent the trickle. Substituting a 12AT7 for the 12AX7 in the phase inverter, for example, comes at minimal cost, so the Deluxe increases the transconductance[132] of its long-tailed pair right alongside its higher-priced siblings. A similar argument can be made for the blackface tremolo upgrade – without it the required extra triode goes unused. The Deluxe thus gets signal modulation during the same design iteration as seven other blackface amps.

The AA763 Deluxe implements a long list of upgrades, some at once and others, ahem, at last. (The DC voltages shown in the schematics on the next two pages are Fender's measurements.)

---

[129] "Joe's Guitars" Available at https://jbonamassa.com/guitars/ (Retrieved January 3, 2021)
[130] David Hope and Julian Limberg, "The Economic Consequences of Major Tax Cuts for the Rich," International Inequalities Institute Working Papers, London School of Economics and Political Science, December 16, 2020.
[131] Tom Wheeler, **The Soul of Tone**, (Milwaukee: Hal Leonard, 2007), p. 150.
[132] Richard Kuehnel, **Guitar Amplifier Electronics: Basic Theory**, (Seattle: Amp Books, 2018), p-. 40-43.

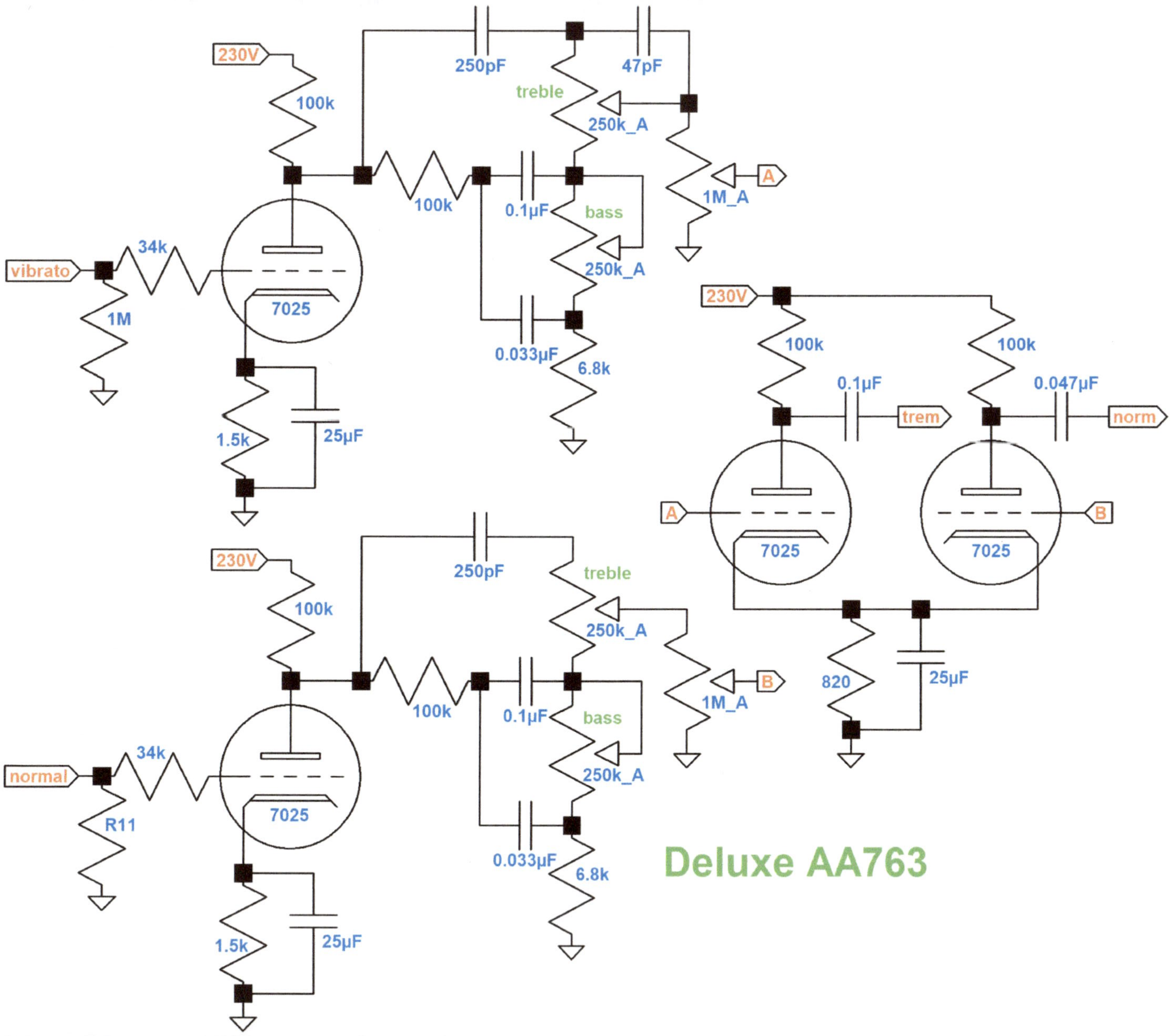


Deluxe AA763

# Deluxe AA763

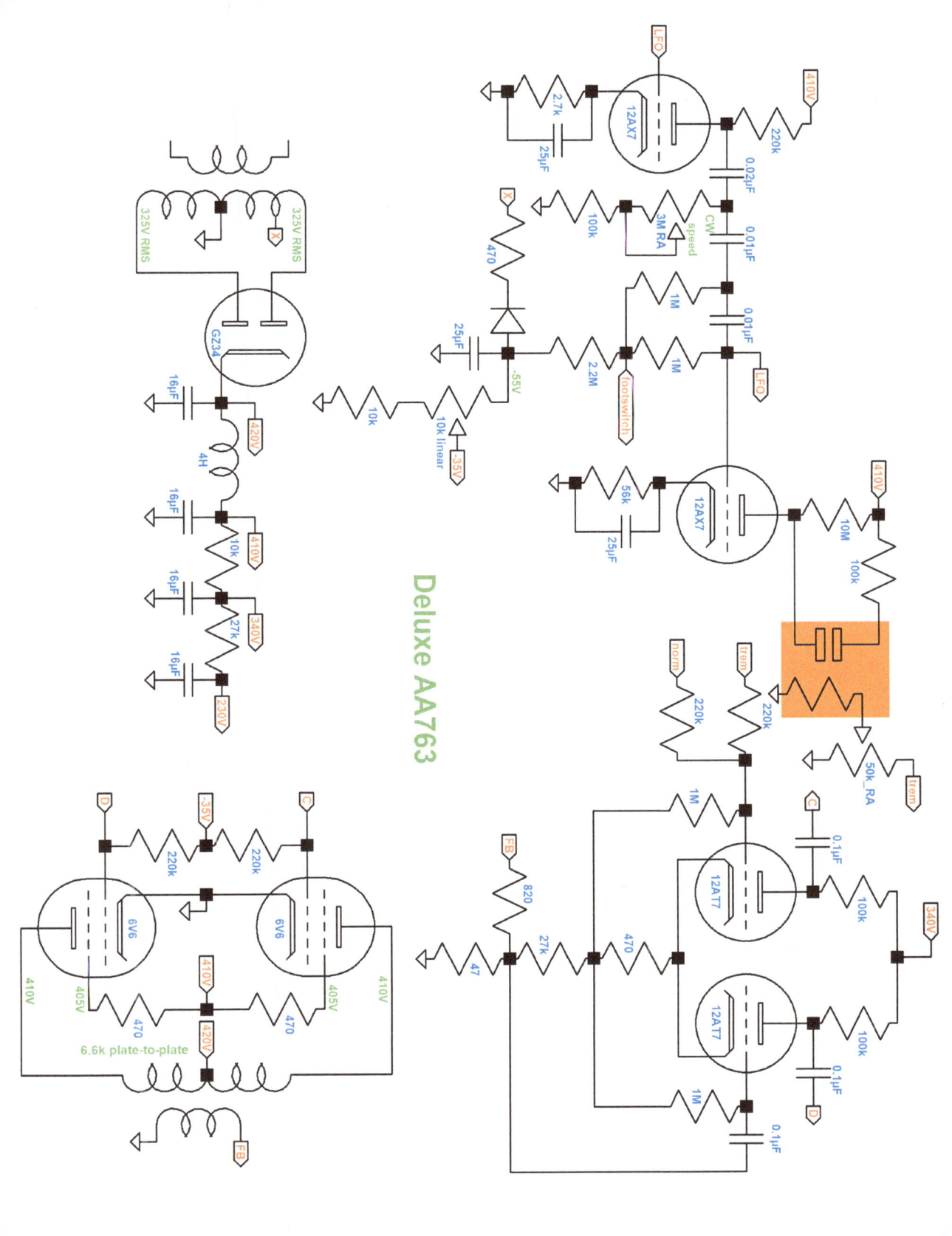

**tremolo** — Tremolo moves from bias modulation of the power amp to signal modulation in the preamp using a neon lamp and a light-dependent resistor (LDR).

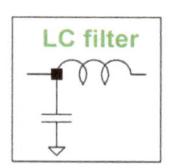

**LC filter** — A choke replaces the first ripple filter resistor to increase filtering without increasing the DC voltage drop across it.

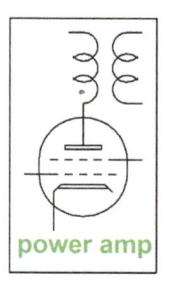

**power amp** — Higher supply voltages drive the 6V6 power tubes to greater output power. Screen resistors are added to prevent radiofrequency oscillation. DC grid bias is more negative and is now adjustable.

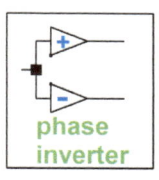

**phase inverter** — A 12AT7 replaces the 12AX7 for the long-tailed pair. The Concert, Pro, Showman, Super, Tremolux, Vibrolux, and Vibroverb get the same upgrade.

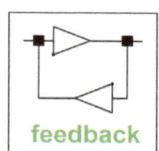

**feedback** — Negative feedback increases between the output transformer secondary and the long-tailed pair.

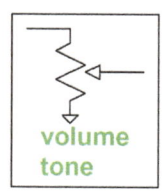

**volume tone** — Each channel gets a tone stack with bass and treble controls that add scooped mids to system voicing. A bright bypass capacitor is added to the vibrato channel's volume control.

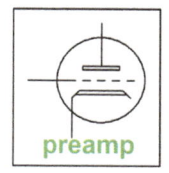

**preamp** — Each channel gets two voltage amplification stages in front of the phase inverter.

## Coaxing Even More Power from the Power Amp

"With the change from the tweed circuits to the browns and blondes and then the blackfaces, there was an appreciable increase in power-supply voltages, resulting in a more powerful, brighter, stronger, and fuller sound with greater punch and frequency response. This went hand in hand with the wider-band speakers that were becoming available at the same time."[133] —Greg Huntington

Power amp supply voltages are increased and 470Ω screen resistors are added to prevent radiofrequency oscillation.

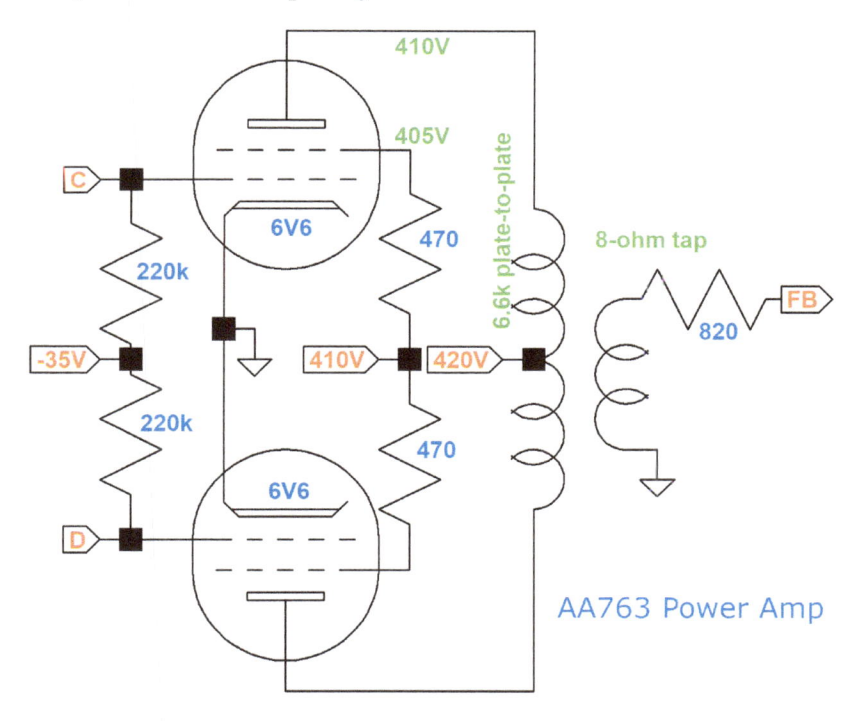

*AA763 Power Amp*

Fender measures 410V for the screen supply and 405V at the screens, representing a 5V drop across each 470Ω screen resistor. According to Ohm's Law, DC screen current is

$$\frac{5V}{470\Omega} = 10.6 mA$$

---

[133] Tom Wheeler, **The Soul of Tone**, (Milwaukee: Hal Leonard, 2007), p. 263.

This is an intense electron flow for a Class AB amp at idle. Screen dissipation for a 405V screen is more than double the 6V6 limit of 2W. The exact DC screen current value is not critical for a guitar amp, but it is likely closer to 2mA per tube. This creates only a 1V drop across the screen resistors for a screen voltage of 409V. A 1V difference is only a tenth of a percent for a 1000V range, a tolerance that expects too much from a circa 1960 voltmeter. Fender's schematic states that its measurements are plus or minus 20 percent.

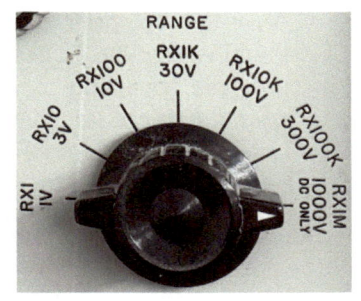

The 6V6 plates and screens are nearly the same voltage. This means we can use the data sheet's triode-connected plate characteristics to determine the cathode current, i.e. the sum of plate and screen current. Since the screen has the most effect on plate current, we use the screen voltage for both electrodes. The red dot indicates that for a -35V grid and 409V "plate" the "plate current" is about 30mA.

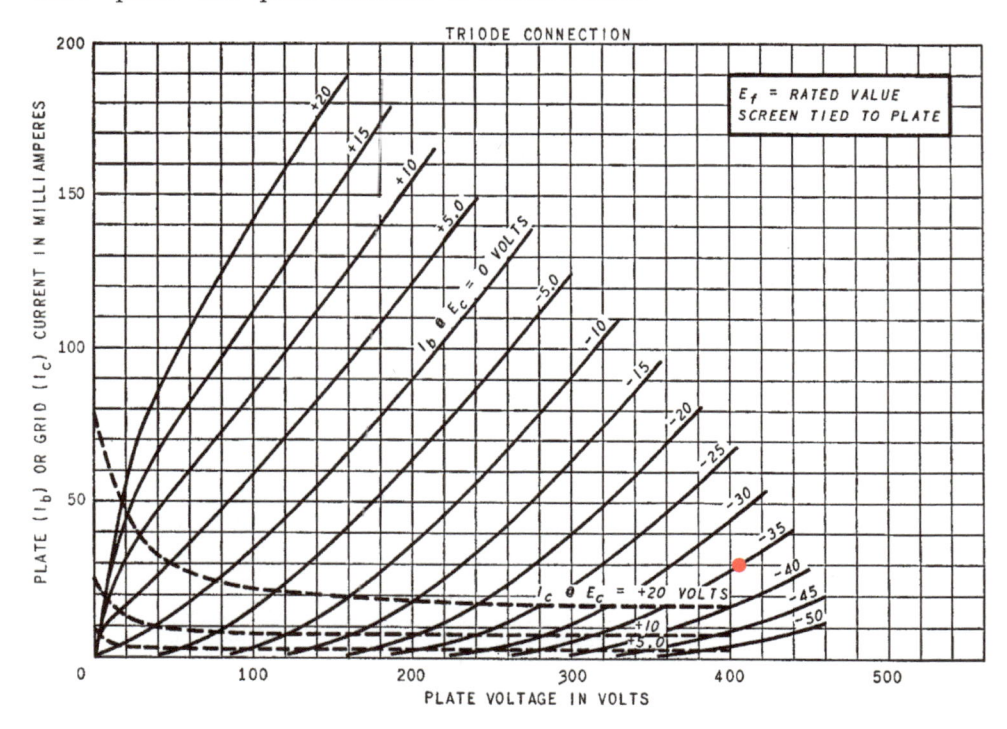

## Screen Voltage Swing at Full Power

Here are estimated 6V6 plate characteristics for screen voltages from 250V to 400V for a 0V grid.

146 – CHAPTER 5

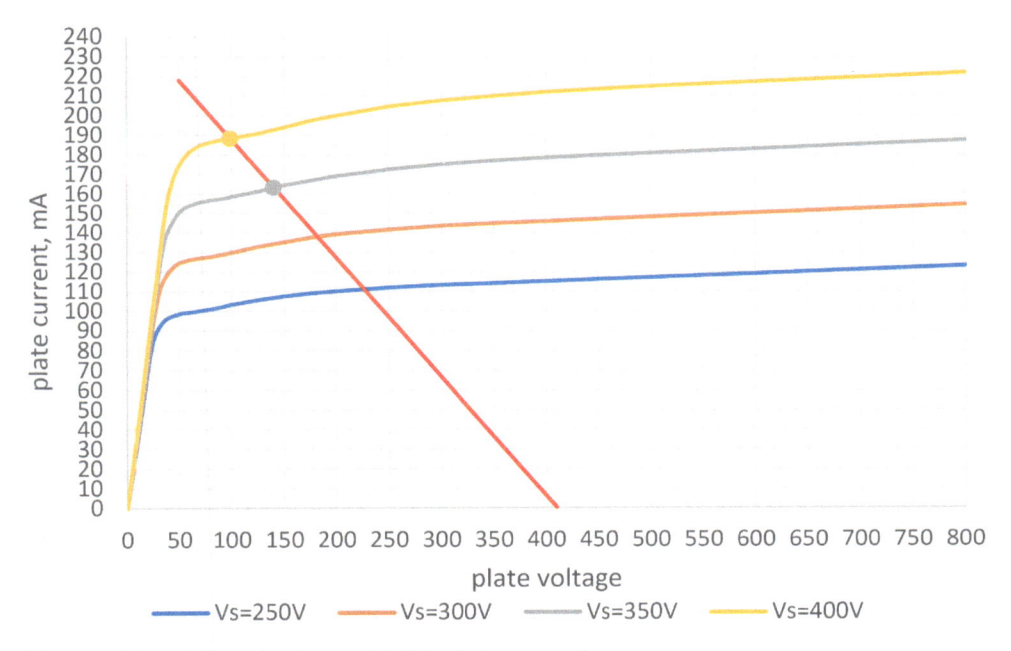

The red load line is for a 410V plate supply and a 1.65kΩ plate load (6.6kΩ plate-to-plate). At idle each tube sees -35V at the grid and 409V at each screen. At full power, as a grid swings to 0V, screen current increases, which increases the voltage drop across the 470Ω screen resistor, reducing the screen voltage. If the screen voltage sags to 400V, for example, then the load line intercepts the yellow 400V curve, as shown by the yellow dot, where the plate voltage is 99V. According to Ohm's Law, for a 410V screen supply and a 400V screen, screen current is

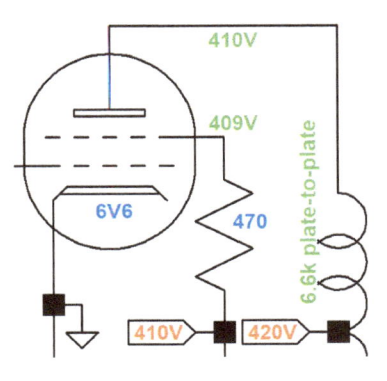

$$\frac{410V - 400V}{470\Omega} = 21.2mA$$

The gray dot indicates that if the screen voltage swings to 350V at full power then the plate swings to 140V and screen current is

$$\frac{410V - 350V}{470\Omega} = 128mA$$

Here are estimated curves for screen current with a 0V grid and four different screen voltages.

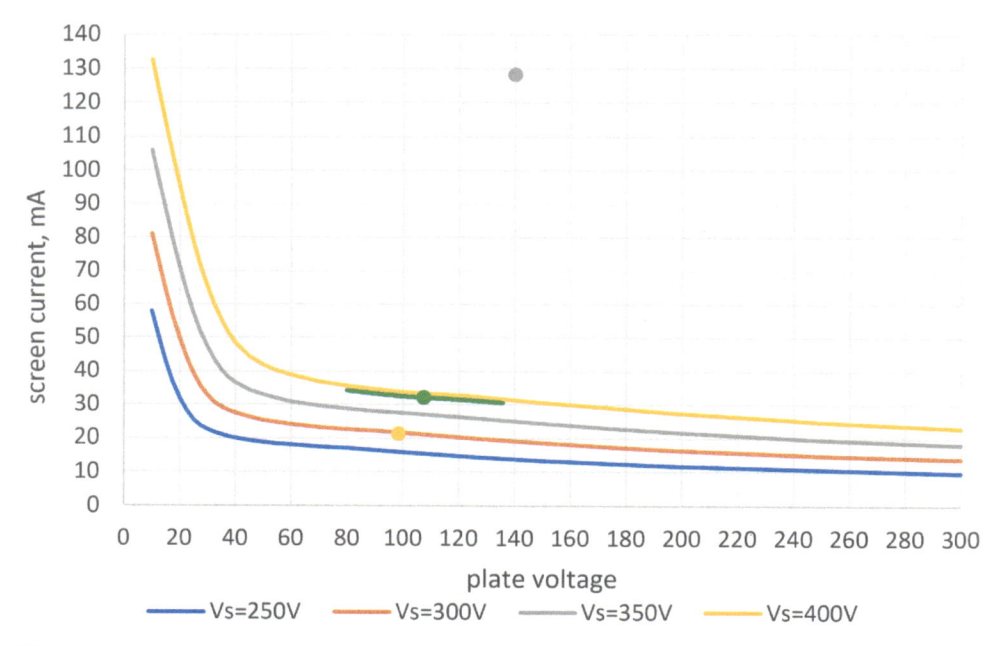

The yellow dot is for a 99V plate and 21.2mA screen current. The gray dot is for 140V, 128mA. Neither dot agrees with its corresponding curve of the same color. The yellow dot is closer to the yellow curve, so we can expect our estimate of the screen voltage at maximum swing to be closer to 400V than to 350V. In fact, a 395V screen, a screen current of 32mA, and a 108V plate seem to work well, as shown by the green dot and an estimated green curve for a 395V screen.

Output power decreases somewhat due to the screen voltage sag. Without the screen resistors, the screen voltage would be 410V and plate voltage would swing to an imagined 410V curve on the plate characteristics shown on the next page. With the screen resistors, the plate voltage swings from its DC value of 410V to the green dot on an estimated 395V curve: 108V, 184mA. Approximate power is

$$\left(\frac{1}{2}\right)(410V - 108V)(184mA) = 27.8W$$

This represents 1dB more power than for the 6G3 design. Like for our previous estimates of output power, this estimate assumes the power supply voltages do not sag at full power, which of course they do, particularly when the power amp is biased for Class AB operation.

The DC grid bias is -35V, so input headroom is 35V peak (+27.9dBV).

— Vs=250V    — Vs=300V    — Vs=350V    — Vs=400V

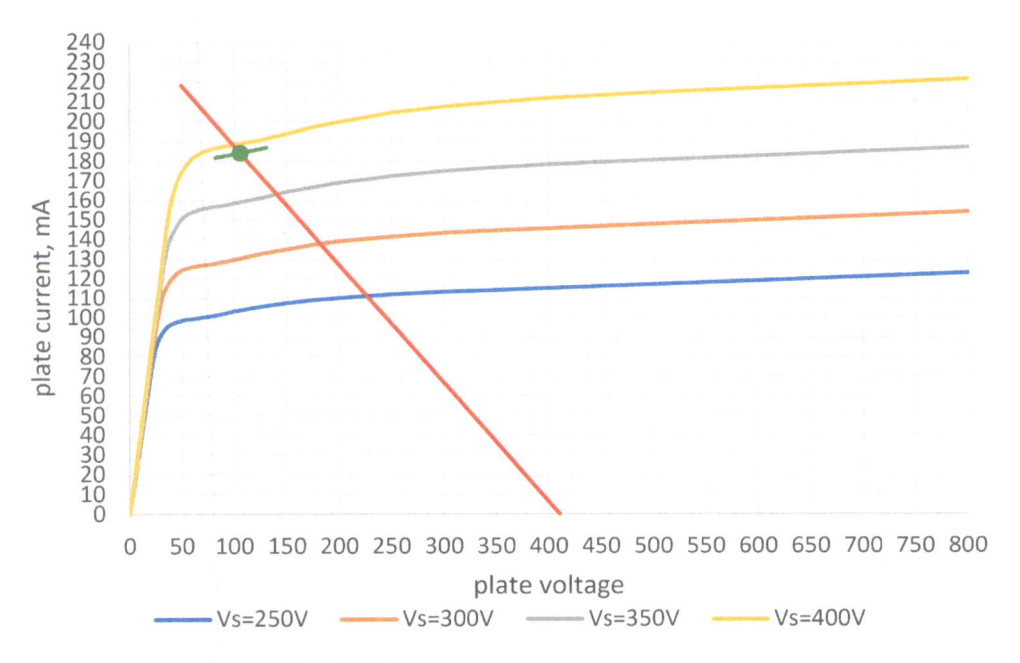

## Power Supply Modifications

The power transformer has a dedicated tap in the secondary for the DC grid bias supply, as shown here by node X.

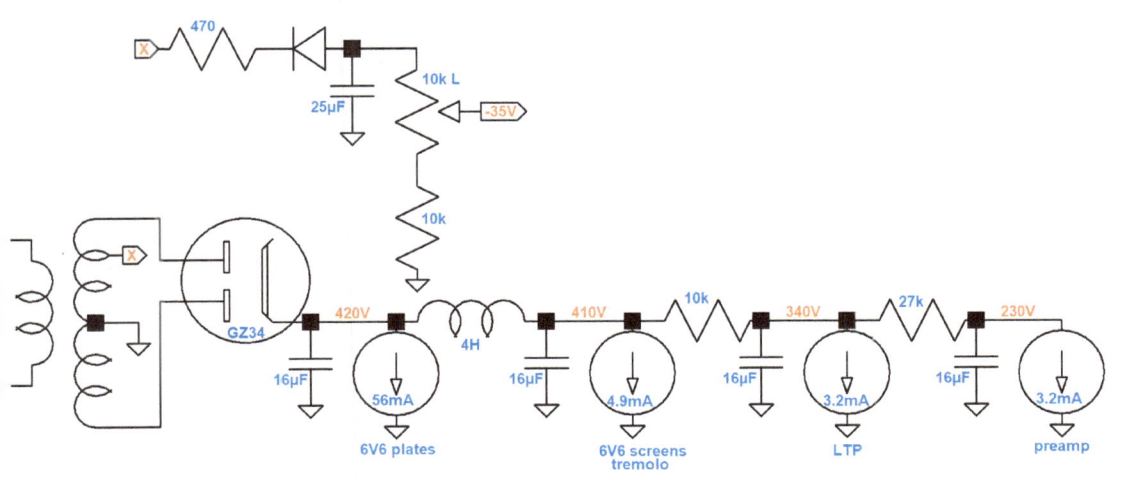

The power supply also includes a choke. Its low DC resistance increases the 6V6 screen supply voltage by 45V.

Based on our estimate of the power amp DC operating point, the power tubes draw 60mA at idle. The triodes draw 7.3mA (as we will determine) for a total power supply load of 67.3mA. This is less than for the 6G3, whose power tube grids are biased more warmly at -26V. For a DC load of 67.3mA at 420V, a 16µF capacitor input filter, and a ripple frequency of 120Hz, the ratio of the ripple voltage in RMS to DC voltage is[134]

$$\frac{\sqrt{2}(67.3mA)}{2\pi(120Hz)(420V)(16\mu F)} = 0.0203$$

This means the ripple level is $(0.0203)(420V) = 8.5V\ RMS$, which is 12V peak (+18.6dBV).

Ripple attenuation for the LC filter with a 4H choke and 16µF capacitor is

$$\left|\frac{1}{-[2\pi(120Hz)]^2(4H)(16\mu F) + 1}\right| = 0.0283\ (-31dB)$$

AC ripple at the 6V6 screens is therefore

$$18.6dBV - 31dB = -12.4dBV\ (339mV\ peak)$$

Ripple attenuation for the first RC filter, with a 10kΩ resistor and 16µF capacitor, is

$$\frac{1}{\sqrt{[2\pi(120Hz)(10k\Omega)(16\mu F)]^2 + 1}} = 0.00829\ (-41.6dB)$$

AC ripple for the phase inverter is

$$-12.4dBV - 41.6dBV = -54dBV\ (11mV\ peak)$$

Fender doubles the final filter capacitor value to 16µF. Ripple attenuation with a 27kΩ resistor is

$$\frac{1}{\sqrt{[2\pi(120Hz)(27k\Omega)(16\mu F)]^2 + 1}} = 0.0031\ (-50.3dB)$$

AC ripple for the preamps is therefore

$$-54dBV - 50.3dBV = -104.3dBV\ (8.6\mu V\ peak)$$

## Introducing Signal-Modulation Tremolo

The new tremolo design uses a neon lamp combined with a light-dependent resistor (LDR). The lamp and LDR are optically coupled but electrically isolated from each other: a voltage across the lamp illuminates the LDR to change its resistance. The first triode shown on the next page is a low-frequency oscillator (LFO) similar to the one in the 6G3. When node `footswitch` is disconnected from ground, -55V is applied to both grids via node `in`. This puts the triodes deep into cutoff, deactivating the

---

[134] **Silicon Rectifier Handbook**, (Bloomington: Sarkes Tarzian, 1960), pp. 41, 42.

oscillator and lamp. When the `footswitch` node is grounded, the LFO grid gets a sudden jolt to 0V, kick-starting the oscillator and activating the lamp.

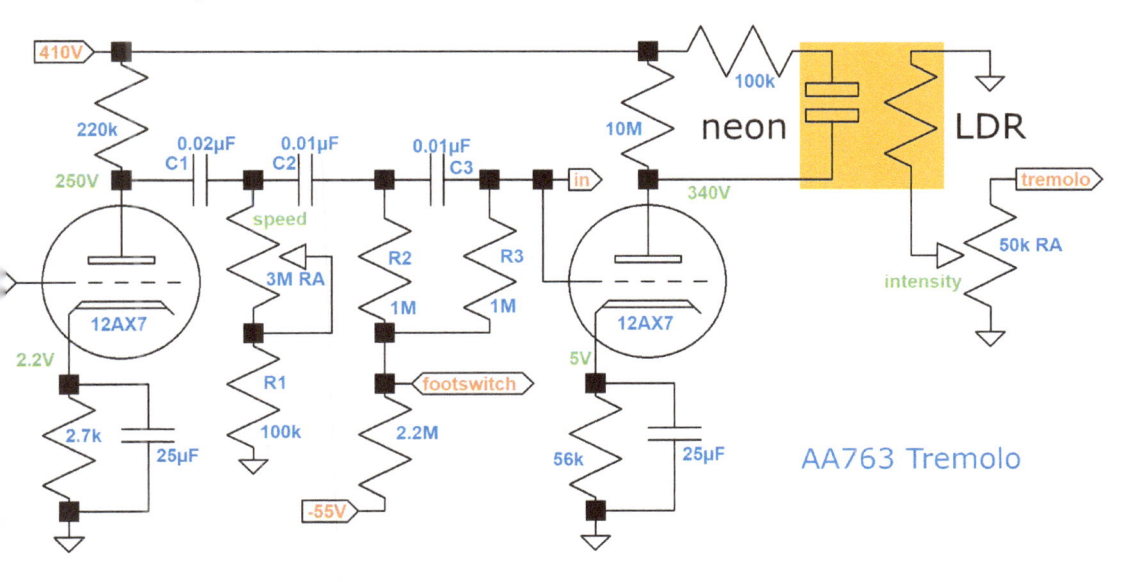

*AA763 Tremolo*

The second triode acts as a buffer that drives the neon lamp based on the LFO signal. The amount of current through the lamp is controlled by the triode circuit, as we will examine shortly. The 10MΩ resistor keeps a tiny amount of plate current flowing when the lamp is off. When the lamp shines brighter, LDR resistance decreases, which shunts more of the guitar signal to ground at the vibrato channel's second-stage output.

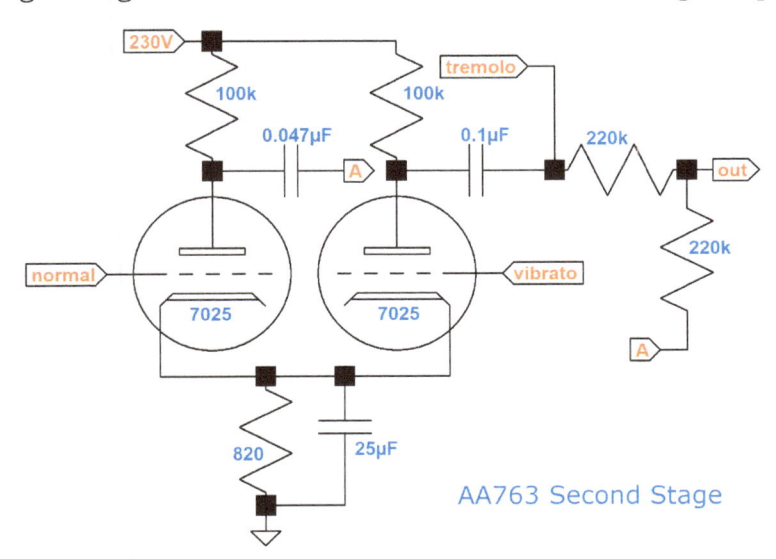

*AA763 Second Stage*

**PERFECTION – THE BLACKFACE DELUXE REVERB – 151**

In this way a variation in light intensity changes the guitar signal level.

An audio taper is at 10-percent resistance when the knob is at 50-percent rotation. The "intensity" potentiometer has a *reverse audio taper*. At 50-percent rotation it is at 90-percent resistance.

When the footswitch is closed, the LFO triode's grid is at 0VDC and we can draw a traditional DC load line and grid line.

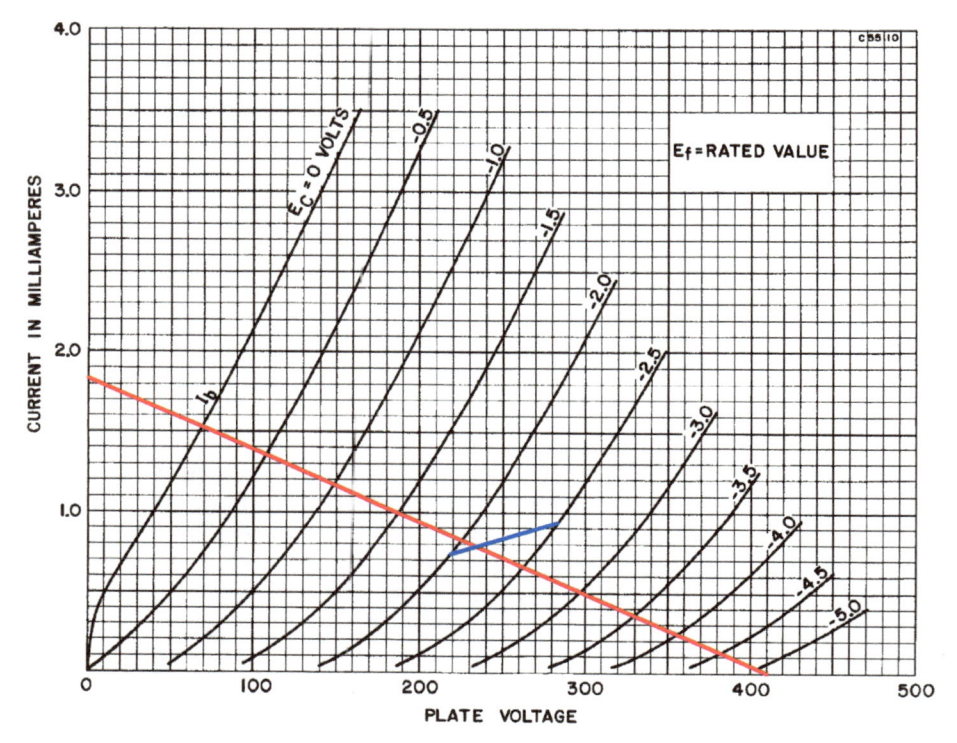

The DC operating point is defined by a -2.1V DC grid bias (Fender measures -2.2V), 234V plate-to-cathode (Fender measures 160V), and 0.8mA plate current.

For the buffer circuit, Fender measures 5VDC across the 56kΩ cathode resistor for a plate current of

$$\frac{5V}{56k\Omega} = 0.089 mA \ (89 \mu A)$$

There is 70V across the 10MΩ resistor for a current of only

$$\frac{70V}{10M\Omega} = 7\mu A$$

The remainder of the plate current, 82µA, flows through the neon lamp and 100kΩ ballast resistor. This puts the lamp in the *normal glow* region of operation, as shown here for a generic lamp.[135]

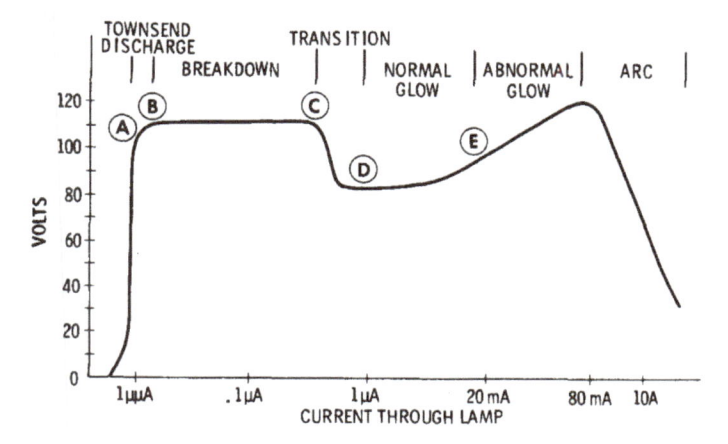

The X axis represents current through the lamp using a logarithmic scale. The Y axis is the voltage across the lamp using a linear scale. A non-conducting lamp begins to conduct when its breakdown voltage is reached (point B). As the current through the lamp increases it transitions (point C) into the normal glow region (between points D and E) in which a lower voltage, called *maintaining voltage*, is sufficient to maintain current flow.

In the normal glow region, the maintaining voltage is fairly constant over a wide current range,[136] so the lamp acts like a fixed voltage drop in the triode plate circuit. Fender measures 70V between the plate supply and the plate. The voltage across the 100kΩ ballast resistor is $(82\mu A)(100k\Omega) = 8.2V$, so the maintaining voltage is $70V - 8.2V = 62V$.

Circuit operation is the same if we reverse the position of the ballast resistor and lamp. Moreover, we can ignore the 10MΩ resistor because it conducts only a tiny amount of current, effectively operating as an open circuit for normal glow conditions. An equivalent circuit therefore drops the plate supply voltage by 62V for an effective supply of 348V, as shown here.

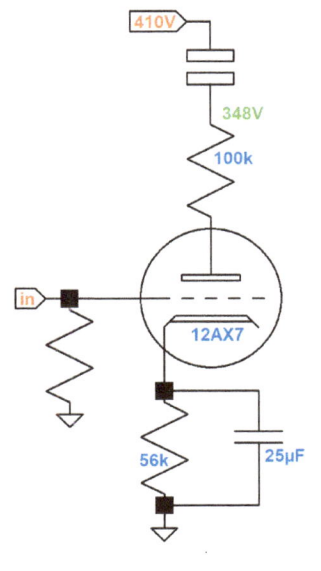

---

[135] William Miller, **Using and Understanding Miniature Neon Lamps**, (New York: Howard W. Sams, 1969), p. 8.
[136] William Miller, p. 7.

The DC operating point can be determined in a conventional way for a 348V plate supply, 100kΩ plate load resistor, and 56kΩ cathode resistor.

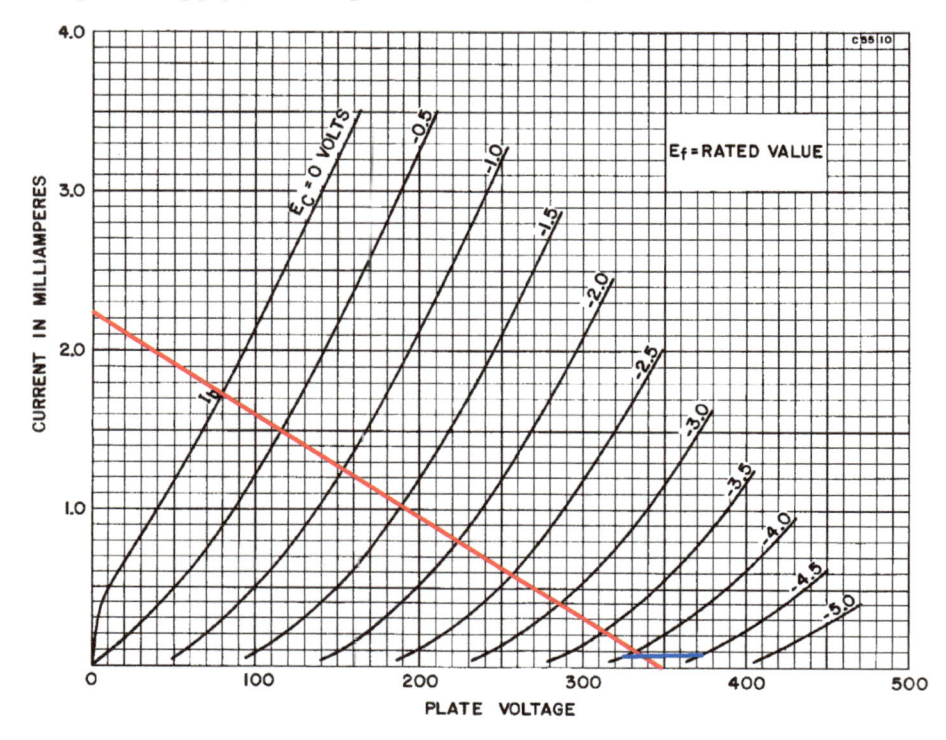

The intersection of the red DC load line and blue grid line indicates that the DC operating point is defined by a DC grid bias of -4.2V (Fender measures -5V), 336V plate-to-cathode (Fender measures 335V), and 82μA plate current.

The AC plate load is the same as for DC: 100kΩ, so the AC load line matches the DC load line. In this icy cold operating region near cutoff, it is difficult to precisely predict tube behavior, but it appears that a 500mV increase in grid voltage creates a 150μA increase in lamp current and a corresponding increase in the illumination of the LDR.

## Modifying the LTP Phase Inverter

The AA763 replaces the phase inverter's 12AX7 dual triode with a 12AT7.

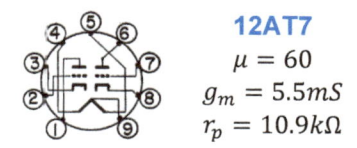

**12AT7**
$\mu = 60$
$g_m = 5.5mS$
$r_p = 10.9k\Omega$

## AA763 Phase Inverter

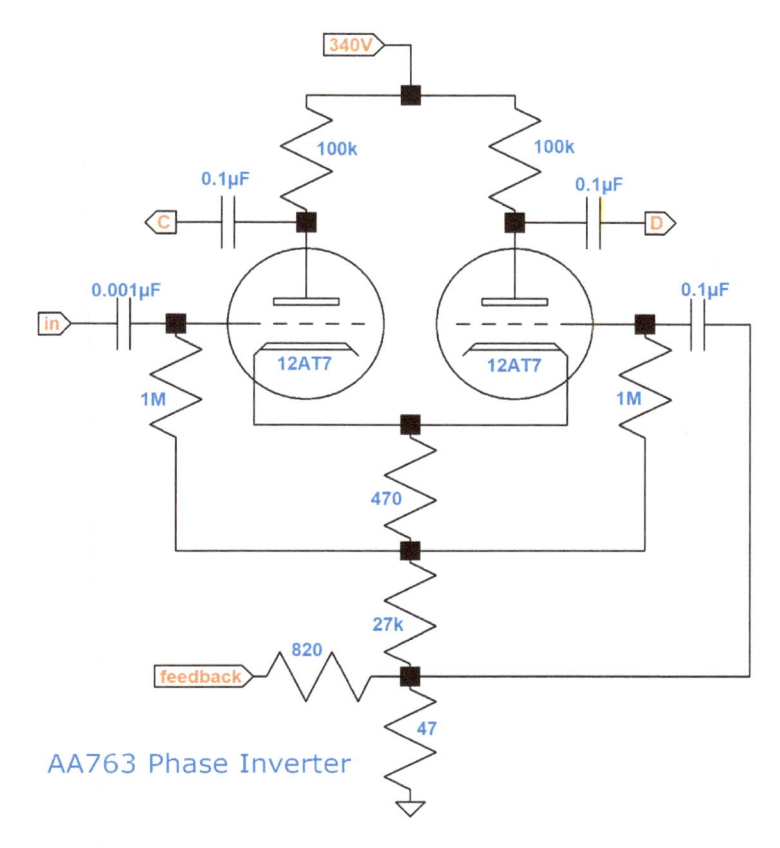

At DC the capacitors are open circuits. The resistors between the cathodes and ground carry the current of two triodes, so the equivalent resistances for one triode are doubled: 940Ω, 54kΩ, and 94Ω. The output transformer secondary has a very low DC resistance to ground, so the 820Ω feedback resistor is in parallel with the 47Ω resistor. The latter, however, is much smaller, so we can ignore the feedback resistor for DC.

On the next page are plate characteristics for a 12AT7. If the plate current is 0mA, the plate-to-cathode voltage is equal to the plate supply voltage: 340V. According to Ohm's Law, if the plate-to-cathode voltage is 0V, the plate current is

$$\frac{340V}{100k\Omega + 940\Omega + 54k\Omega + 94\Omega} = 2.19mA$$

The endpoints of the DC load line are therefore 340V, 0mA and 0V, 2.19mA, as shown by the red line.

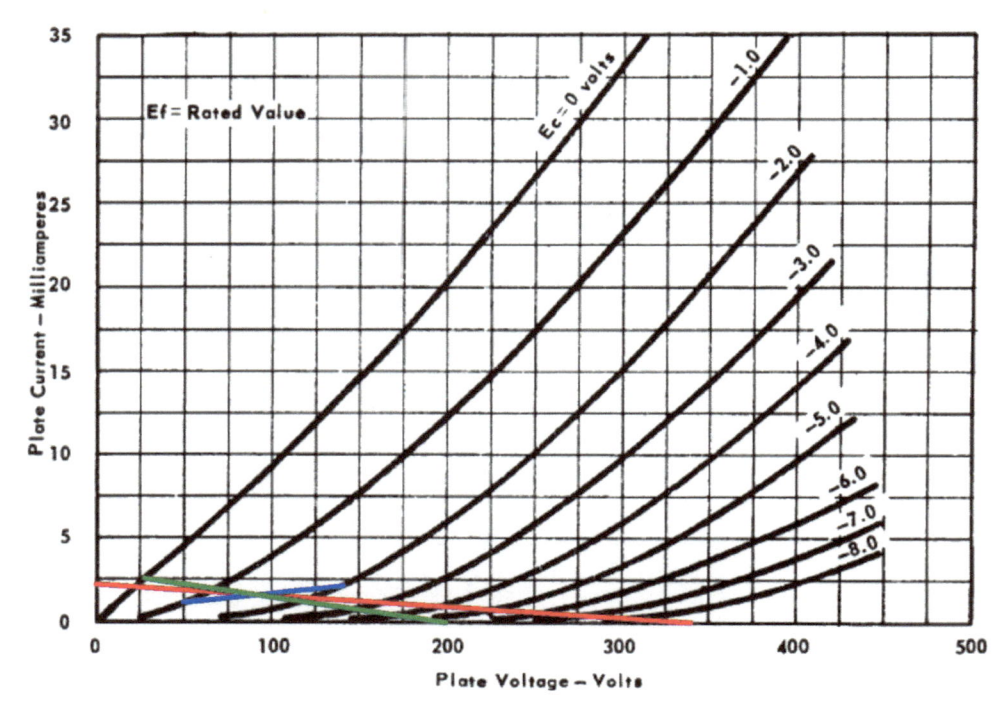

If the grid voltage is -1V or -2V, the plate current is

$$\frac{1V}{940\Omega} = 1.1mA$$

$$\frac{2V}{940\Omega} = 2.1mA$$

These are the endpoints of the blue line. The intersection of the red and blue lines indicates that the DC operating point is defined by a grid-to-cathode voltage of -1.5V, a plate-to-cathode voltage of 90V, and a plate current of 1.6mA. These values match Fender's measurements exactly. (Fender measures a 160V drop across the 100kΩ plate load resistors, corresponding to 1.6mA plate current.)

For audio signals each 100kΩ plate load resistor is in parallel with a 220kΩ grid-leak resistor for an effective AC plate load of 69kΩ. According to Ohm's Law, when the plate current swings from its DC value of 1.6mA to 0mA, the plate voltage swings by $(1.6mA)(69k\Omega) = 110V$. The plate voltage therefore swings from its DC value of 90V to a value of $90V + 110V = 200V$. The AC load line thus includes the DC operating point and the endpoint 200V, 0mA. The resulting AC load line is shown in green.

Grid-to-cathode voltage swing is limited to 1.5V. The grid-to-ground swing is double: 3V, so input headroom is 3V peak (+6.5dBV) without negative feedback, which we examine next.

The plate voltage is limited to a swing of about 65V peak. Input headroom for the power amp is only 35V peak, so the phase inverter can overdrive the power amp by almost a factor of 2 but no further. This is an important limitation – if the phase inverter overdrives the power amp by more than a factor of about 2, *blocking distortion*[137] can occur, which is highly undesirable in a guitar amp.

Loaded voltage gain is about half the voltage gain of a traditional (one triode) voltage amplifier with a 12AT7 amplification factor of 60, a 12AT7 plate resistance of 10.9kΩ, a 69kΩ AC plate load, and a fully bypassed cathode resistor:

$$\left(\frac{1}{2}\right)\left(\frac{(60)(69k\Omega)}{69k\Omega + 10.9k\Omega}\right) = 25.9 \ (28.3dB)$$

This assumes no feedback from the output transformer secondary and that the noninverting grid is connected to AC ground instead of to the top of the 47Ω resistor.

Because the plate load resistor values are the same, the inverted output (node C) has slightly more gain than the non-inverted output (node D). The actual average voltage gain is 25.7 (28.2dB), a value that we use in the next section for negative feedback calculations. The new 12AT7 design has slightly more gain than the 12AX7 version because the tube's lower amplification factor is compensated by a lower plate resistance, which increases its ability to drive the 220kΩ loads.

The power amp needs +27.9dBV (35V peak) at the 6V6 grids to achieve full power. Without global negative feedback, the LTP input needs $27.9dBV - 28.2dB = -0.3dBV$ (1.4V peak).

## Increasing Negative Feedback

The AA763 feedback signal is taken from the same 8Ω tap that drives the speaker. It is attenuated by a voltage divider formed by an 820Ω feedback resistor and the 47Ω resistor at the end of the LTP tail, as shown on the next page. The attenuated signal is then fed to the grid of the triode driving the non-inverted output. This opposes the guitar signal at the opposite grid.

---

[137] Richard Kuehnel, **Guitar Amplifier Electronics: Basic Theory**, (Seattle: Amp Books, 2018), pp. 166-170.

## AA763 Negative Feedback

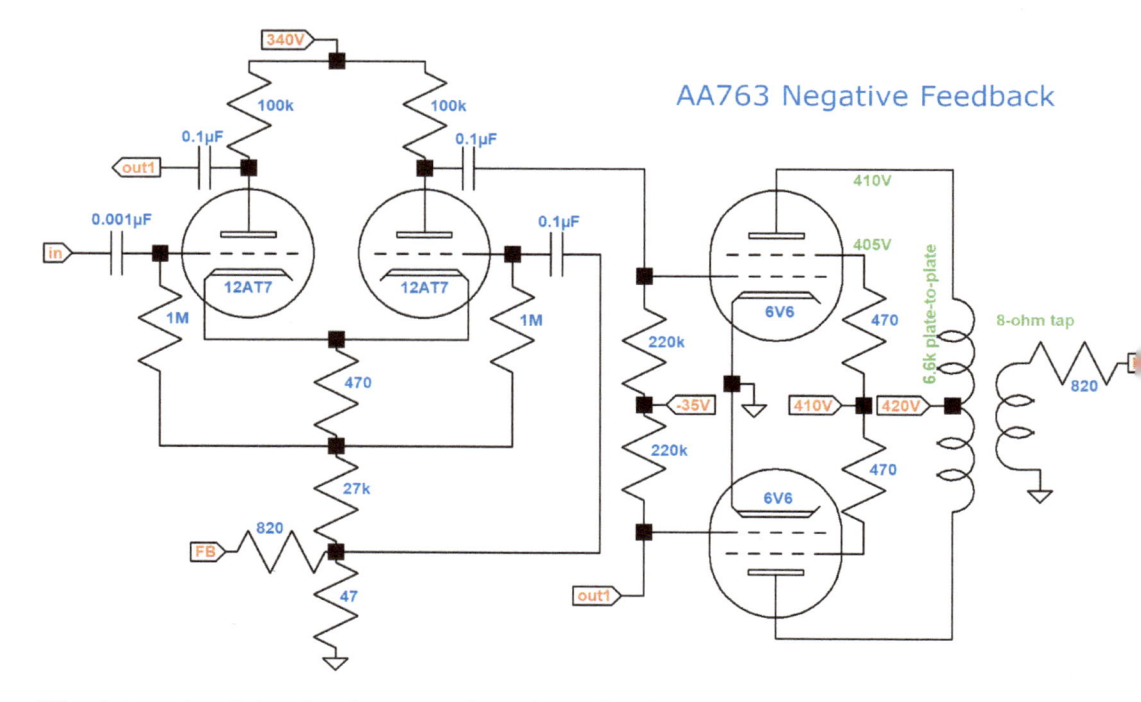

We determined in the last section that the inverted and non-inverted voltage gains for the LTP are $G_1 = G_3 = 25.7$. At full power, the power amp produces about 28W RMS into an 8Ω load (as determined earlier without power supply voltage sag), so the RMS voltage measured at the speaker is

$$\sqrt{(28W)(8\Omega)} = 15V$$

The power tubes are biased at -35V. At full power, the RMS power amp input voltage is

$$\frac{35V}{1.414} = 24.8V$$

The voltage gain for the power tubes and output transformer is the RMS voltage at the speaker divided by the RMS voltage at the power tube grids:

$$G_2 H_1 = \frac{15V}{24.8V} = 0.6$$

The phase inverter feedback voltage gain from the transformer secondary to the LTP feedback input is approximately the gain of the voltage divider formed by the 820Ω feedback resistor and the 47Ω resistor connecting it to ground:

$$H_2 = \frac{47\Omega}{820\Omega + 47\Omega} = 0.054$$

158 – CHAPTER 5

Now that all the individual gains are known, we just crunch the numbers. The forward (open-loop) voltage gain is

$$G = G_1(G_2H_1) = (25.7)(0.6) = 15.4 \ (23.8dB)$$

The feedback gain is

$$H = H_2\frac{G_3}{G_1} = (0.054)\left(\frac{25.7}{25.7}\right) = 0.054$$

The closed-loop voltage gain, according to the famous feedback formula, is

$$G_{CL} = \frac{G}{1+HG} = \frac{15.4}{1+(0.054)(15.4)} = 8.4 \ (18.5dB)$$

The voltage gain with negative feedback is 55 percent of the voltage gain without feedback, i.e. 5.3dB less. This is 2.1dB more attenuation compared to the 6G3, but still only a modest amount, especially compared to a high-fidelity amplifier.

> "Another general trend: over time, Leo tended to increase the amount of feedback in his amps – no feedback on some of the tweeds, moderate amounts on the brown amps, and slightly more on the blackface and silverface amps. This decreased distortion and increased bandwidth, for more of a 'hi-fi' sound in the post-tweed Fenders compared to the grungy, roots-blues 'dirt' you hear in most tweed amps at higher volumes."[138] –Tom Wheeler

Phase inverter gain is 28.2dB without negative feedback. With negative feedback it is 5.3dB less. The LTP needs -0.3dBV at its input to achieve full power without feedback, so with feedback, it requires $-0.3dBV + 5.3dB = +5dBV$ (2.5V peak).

We determined in the last section that input headroom is +6.5dBV without feedback. With feedback it is $6.5dBV + 5.3dB = +11.8dBV$ (5.5V peak).

The large 0.1µF coupling capacitors and negative feedback for the 6G3 create a very flat frequency response. Increased feedback in the AA763 flattens the response even further. Feedback also reduces nonlinear distortion from the power amp and lowers the power amp's output impedance, making it more capable of handling the frequency-dependent nature of loudspeaker impedance, particularly for bass.

Increased feedback creates a snappier transition to overdrive and back, as described in the previous chapter. This gives the blackface Deluxe more of

---

[138] Tom Wheeler, **The Soul of Tone**, (Milwaukee: Hal Leonard, 2007), pp. 263-264.

an "on the edge"[139] dynamic.

## Second-Stage Voltage Amplifiers

Each channel has a 7025/12AX7 second-stage voltage amplifier with outputs mixed using 220kΩ resistors.

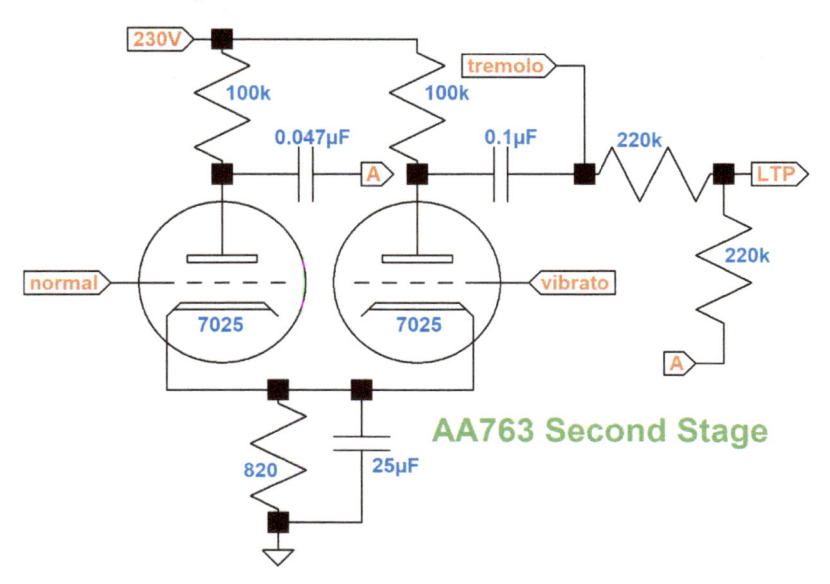

**AA763 Second Stage**

The design is similar to the first-stage preamps of the 6G3 but without the volume and tone controls in front of the mixer. The tremolo input connects the signal to ground via a variable resistance synced to the tremolo's low-frequency oscillator.

The equivalent DC circuit for one triode doubles the cathode resistor value to 1.64kΩ. The DC load line is shown on the next page in red. The intersection with the blue grid line marks the DC operating point: -1.3V grid-to-cathode (Fender measures -1.6V), 150V plate-to-cathode (Fender measures 168V), and 0.8mA.

The vibrato channel has the most demanding AC load: the 50kΩ intensity control. For audio it is in parallel with the 100kΩ plate load resistor for an equivalent AC plate load of 33kΩ. When the plate current decreases to 0mA from its 0.8mA DC value, the plate voltage increases by $(0.8mA)(33k\Omega) = 26V$. The plate voltage swings to $150V + 26V = 176V$. This marks an endpoint of the green AC load line.

---

[139] Dave Hunter, **Amped**, (Minneapolis: Voyageur Press, 2012), p. 141.

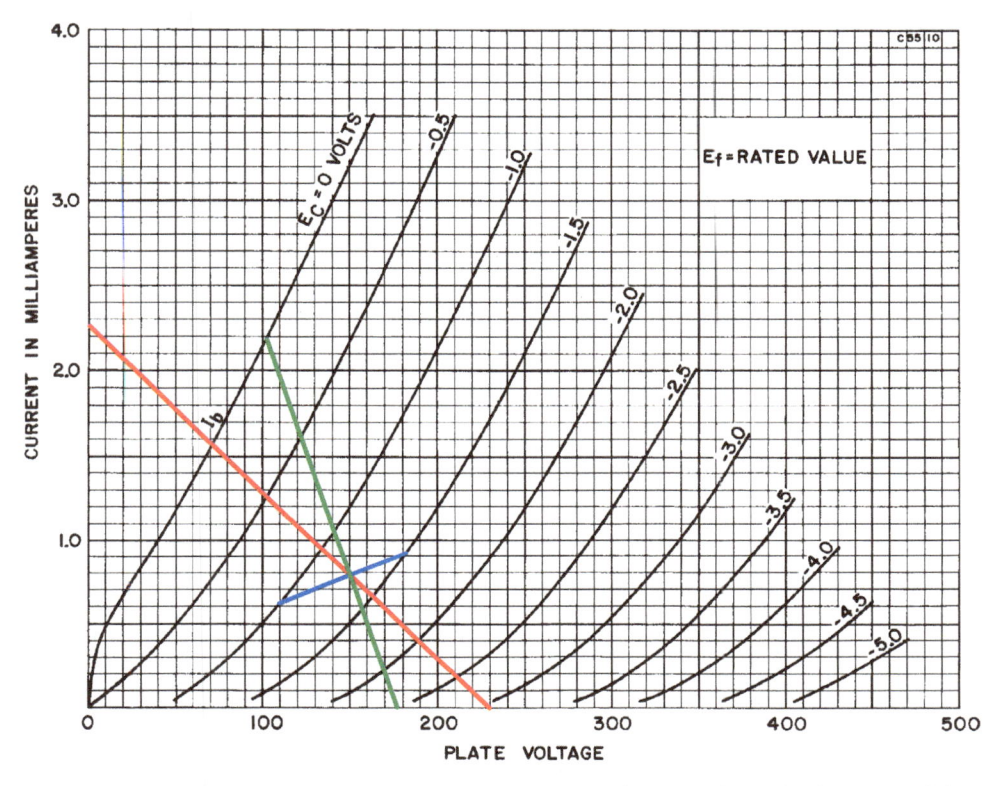

The grid voltage can swing by about 1.2V peak (-1.4dBV) before reaching its limits of headroom.

For the vibrato channel the loaded voltage gain for a 7025/12AX7 amplification factor of 100, a plate resistance of 62.5kΩ, an AC plate load of 33kΩ, and a fully bypassed cathode resistor is

$$\frac{(100)(33k\Omega)}{33k\Omega + 62.5k\Omega} = 34.6\ (30.8dB)$$

The LTP represents a very light load. When the normal channel is inactive, its "signal" is 0V (AC ground) from a voltage amplifier with an output impedance of 38kΩ. Combined with the mixer network, it reduces gain by almost 6dB, so net gain for the vibrato channel from the second-stage grid to the LTP grid is approximately

$$30.8dB - 6dB = 24.8dB$$

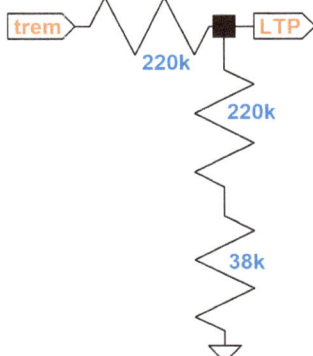

The LTP needs +5dBV at its input to achieve full power, so the second-stage grid needs

$$5dBV - 24.8dB = -19.8dBV\ (145mV\ peak)$$

## First Stage, Tone Stack, and Volume Control

> "A new 'Hi Fidelity' amp with two 12" speakers was unveiled at the summer National Association of Music Merchants (NAMM) show in 1952, featuring increased power and separate Bass and Treble controls. Having the ability to tailor one's sound through the controls of an amplifier, as well as to create a variety of sounds with the same instrument, was a fresh idea."[140] –John Teagle

The first stage is a 7025/12AX7 voltage amplifier with a 100kΩ plate load resistor and a fully bypassed cathode resistor, a circuit that has become ubiquitous for modern guitar amps.

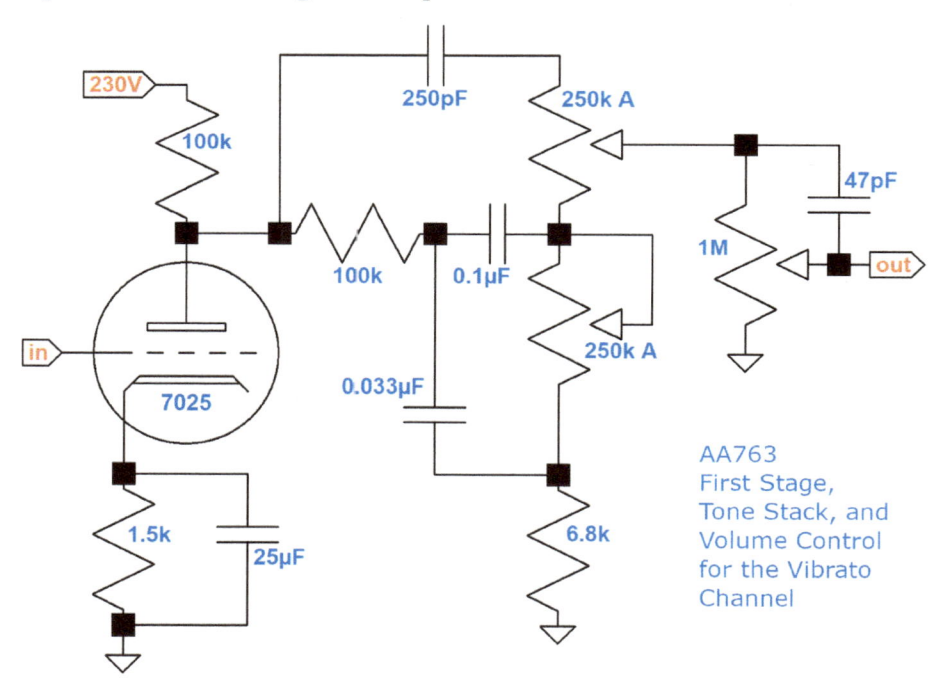

AA763 First Stage, Tone Stack, and Volume Control for the Vibrato Channel

The stack drives a volume control in front of the second stage. For the vibrato channel the control has a 47pF bright bypass capacitor.[141]

---

[140] John Teagle and John Sprung, **Fender Amps: The First Fifty Years**, (Milwaukee: Hal Leonard, 1995), p. 27.
[141] Richard Kuehnel, **Guitar Amplifier Electronics: Basic Theory**, (Seattle: Amp Books, 2018), pp. 72-76.

Here is a SPICE AC analysis simulation[142] of the response from stack input to the second stage grid with the volume control at maximum and the tone controls at minimum and maximum (4 traces total), assuming they are driven by a 38kΩ output impedance (100kΩ plate load resistor and 62.5kΩ plate resistance in parallel).

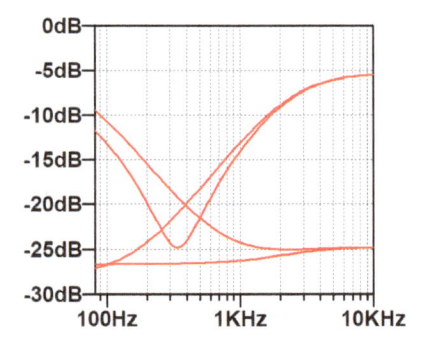

With the controls at maximum middle scoop is substantial. At minimum, the response is flat.

> "I think somebody decided that being able to cut the treble was actually too dark, so they abandoned that in the blackface era. And I think it's fair to say this change is one of the reasons we hear the term 'scooped,' or 'scooped mids' when people describe the tone of the blackface amps compared to earlier Fenders."[143] –Matt Wilkens, Fender engineer

With all controls at 50-percent rotation (10-percent resistance), insertion loss at 1kHz is 43.3dB for the normal channel (left) and 43dB for the vibrato channel (right).

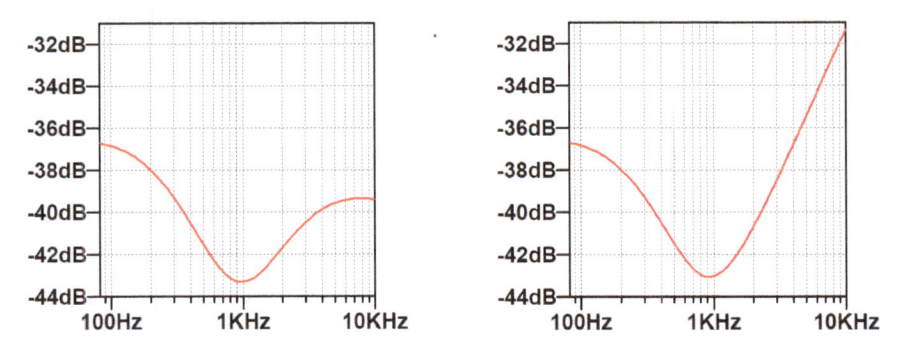

These plots also assume the driving circuit output impedance is 38kΩ.

---

[142] Richard Kuehnel, **Guitar Amplifier Electronics: Circuit Simulation**, (Seattle: Amp Books, 2019).
[143] Tom Wheeler, **The Soul of Tone**, (Milwaukee: Hal Leonard, 2007), p. 267.

A DC load line and grid line indicate that the first-stage DC grid voltage, plate voltage and plate current are -1.2V (Fender measures -1.6V), 147V (Fender measures 168V), and 0.8mA, respectively.

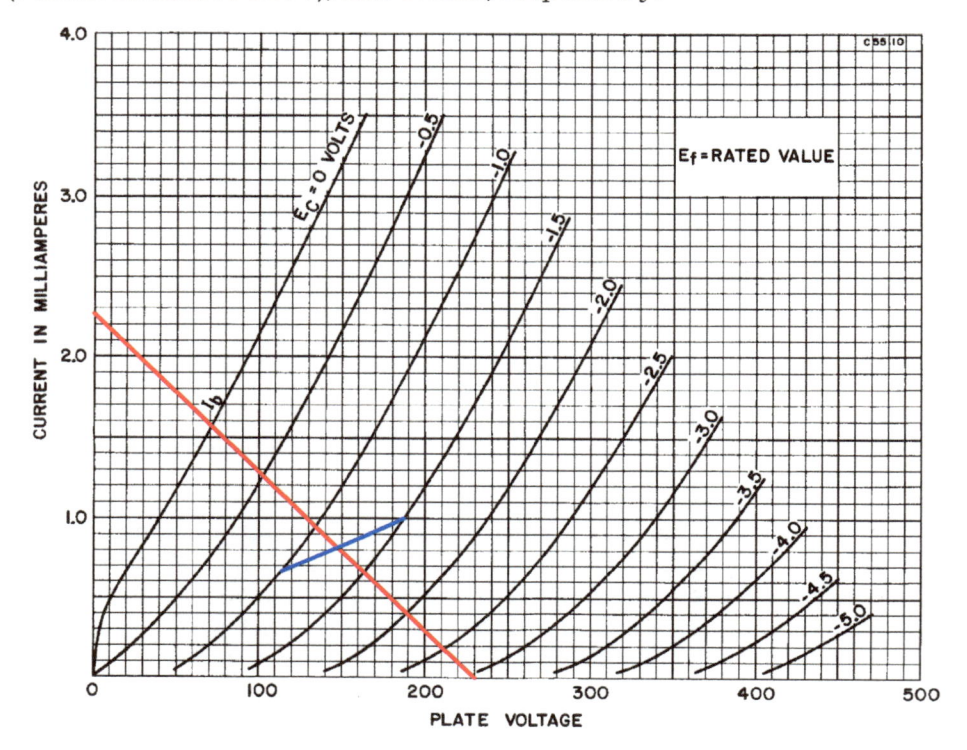

The AC load varies with frequency and knob position. Input headroom is not much of a concern for the first stage, however, so great accuracy is not required. Using the DC load line as an approximation for maximum AC swing, input headroom is about 1.2V peak (-1.4dBV).

The first-stage unloaded gain for a 7025/12AX7 amplification factor of 100, a plate resistance of 62.5kΩ, a 100kΩ plate load resistor, and a fully bypassed cathode resistor is

$$\frac{(100)(100k\Omega)}{100k\Omega + 62.5k\Omega} = 61.5 \; (35.8dB)$$

The second-stage grid needs -19.8dBV for full power. Volume and tone control insertion loss at 1kHz, which accounts for the driving circuit's 38kΩ output impedance, is 43dB (vibrato channel) when the controls are at 50-percent rotation. The guitar signal level at the vibrato input jack needed to drive the power amp to full power with all knobs at 50-percent rotation, is therefore

$$-19.8dBV + 43dB - 35.8dB = -12.6dBV \; (332mV \; peak)$$

164 – CHAPTER 5

With the volume control at maximum, the signal gets almost a 20dB boost for an input sensitivity of -32.6dBV (33mV peak).

## System Profile

The schematics on the next two pages document what we now know about the AA763 when the vibrato channel is active and the amp is operating at full power.

Here is a system profile for the vibrato channel.

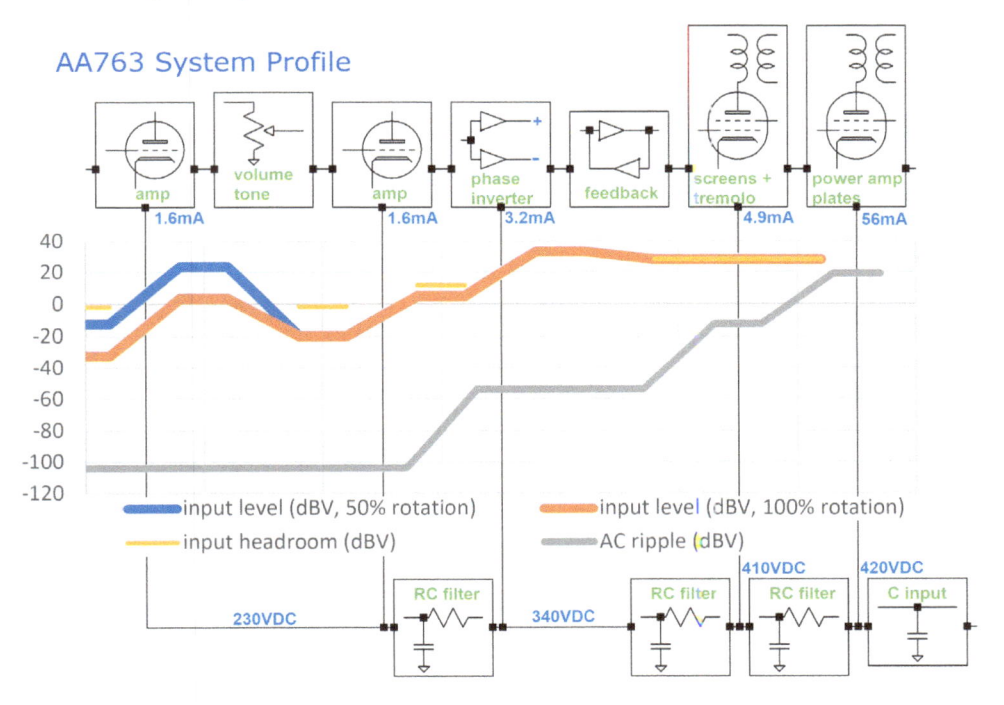

The system profile shows the signal level (blue trace) at the input to each stage at full power with the volume and tone controls at 50-percent rotation. The orange trace is with the volume control at maximum and the tone controls still at noon. The yellow traces show input headroom. The gray trace shows AC ripple for the plate and screen supplies. All traces are in dBV.

Starting at the input jack for the high-gain vibrato channel, the blue trace shows that the guitar signal gets

1. a 35.8dB (unloaded) boost from the first-stage voltage amplifier,
2. 43dB attenuation from the tone and volume control circuit (knobs at noon),

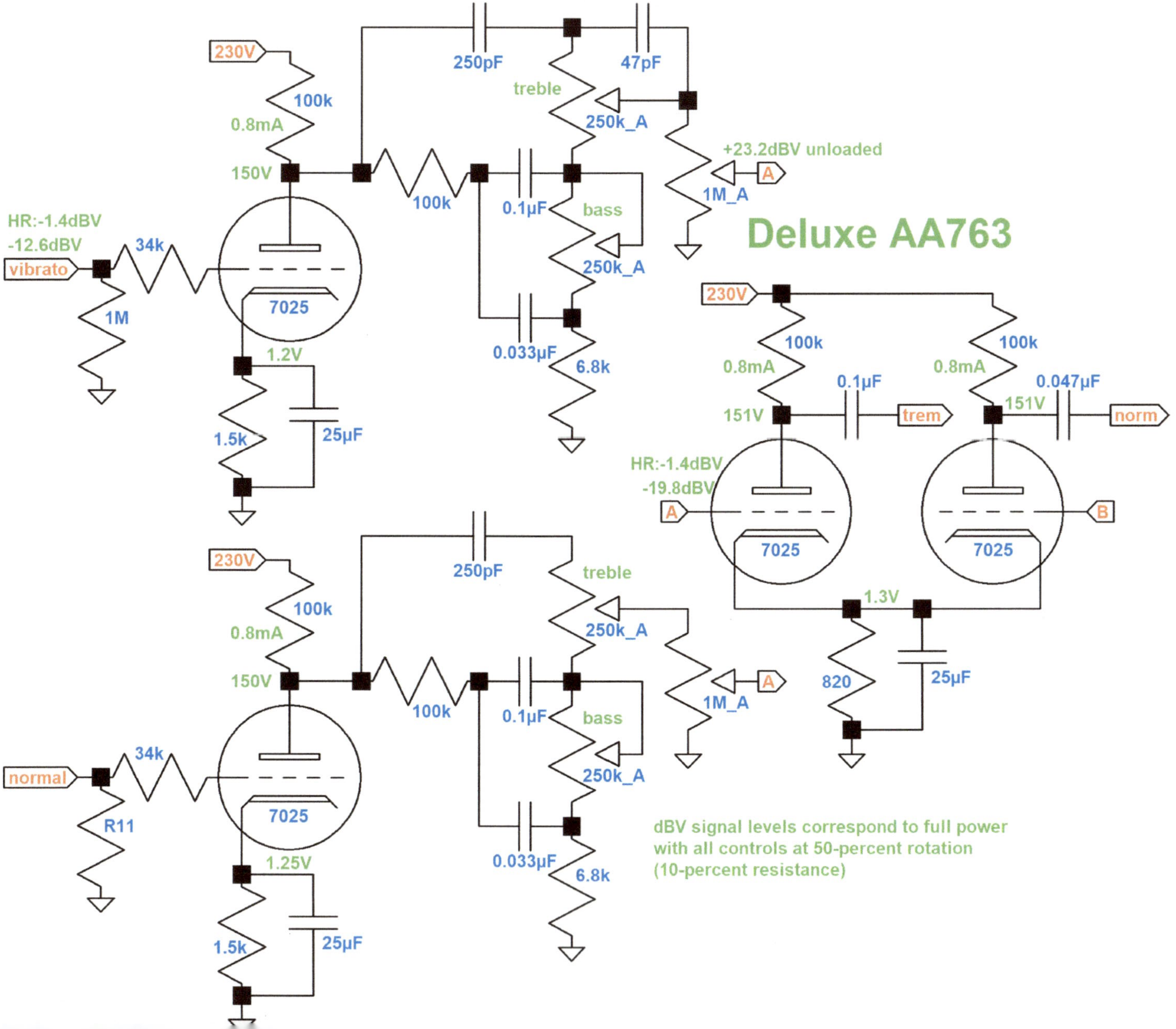

Deluxe AA763
230V
100k
0.8mA
150V
HR:-1.4dBV
-12.6dBV
vibrato
34k
1M
7025
1.2V
1.5k
25µF
250pF
47pF
treble
250k_A
100k
0.1µF
bass
250k_A
0.033µF
6.8k
+23.2dBV unloaded
1M_A
A
230V
100k
0.8mA
151V
0.1µF
trem
100k
0.8mA
151V
0.047µF
norm
HR:-1.4dBV
-19.8dBV
A
7025
7025
B
1.3V
820
25µF
230V
100k
0.8mA
150V
normal
34k
R11
7025
1.25V
1.5k
25µF
250pF
treble
250k_A
100k
0.1µF
bass
250k_A
0.033µF
6.8k
1M_A
A
dBV signal levels correspond to full power
with all controls at 50-percent rotation
(10-percent resistance)

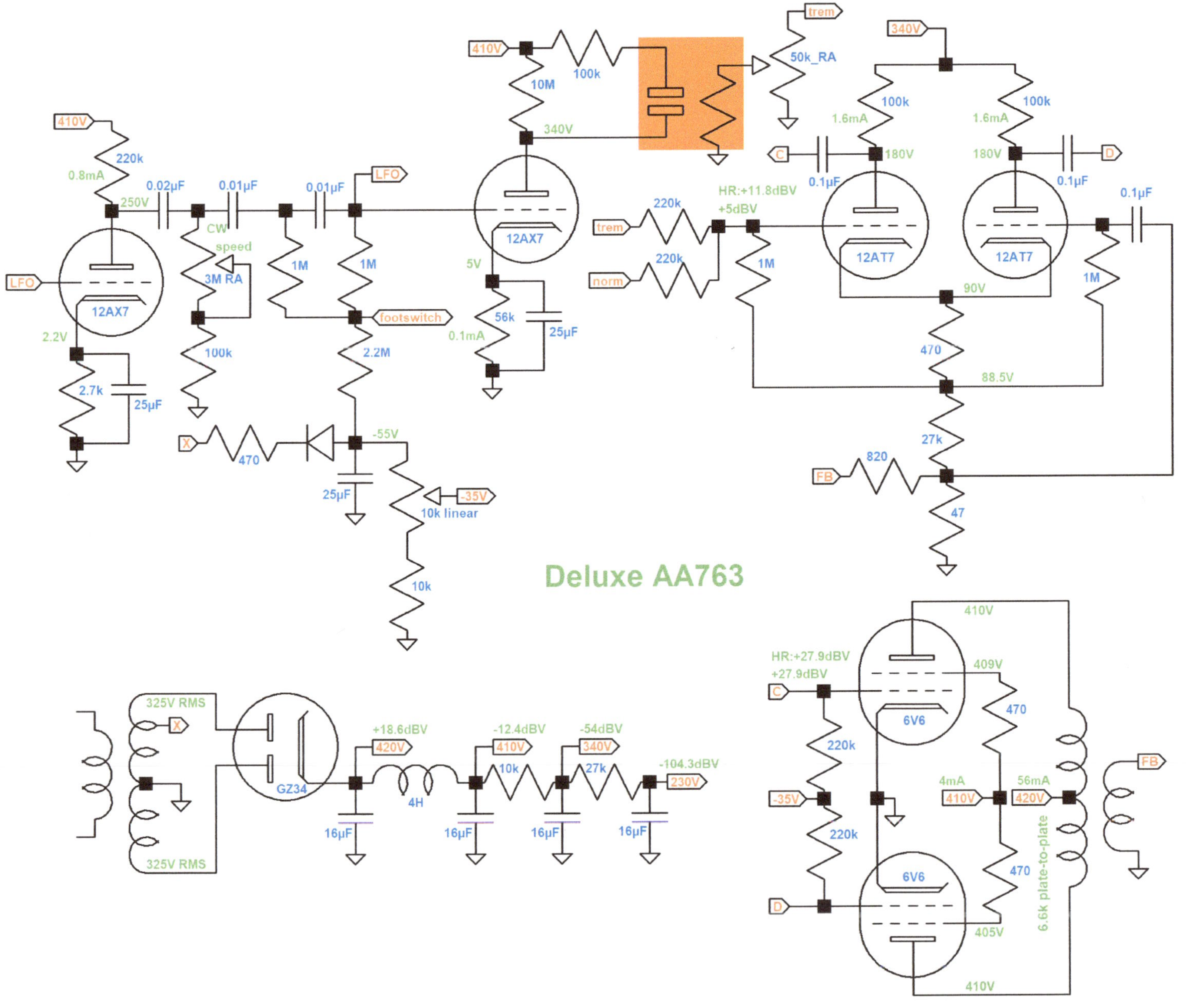
Deluxe AA763
410V
220k
0.8mA
250V
0.02µF
0.01µF
0.01µF
LFO
CW
speed
3M RA
100k
1M
1M
footswitch
2.2M
12AX7
2.2V
2.7k
25µF
X
470
-55V
25µF
-35V
10k linear
10k
410V
100k
10M
340V
12AX7
5V
56k
0.1mA
25µF
trem
50k_RA
340V
100k
100k
1.6mA
1.6mA
C
180V
180V
D
0.1µF
0.1µF
0.1µF
HR:+11.8dBV
+5dBV
trem
220k
norm
220k
1M
12AT7
12AT7
1M
90V
470
88.5V
27k
FB
820
47
325V RMS
325V RMS
GZ34
+18.6dBV
420V
4H
-12.4dBV
410V
10k
-54dBV
340V
27k
-104.3dBV
230V
16µF
16µF
16µF
16µF
410V
HR:+27.9dBV
+27.9dBV
409V
470
6V6
220k
-35V
220k
4mA
410V
56mA
420V
6.6k plate-to-plate
6V6
470
405V
410V

3. a 24.8dB boost from the second-stage voltage amplifier and mixer circuit,
4. a 23.8dB boost from the LTP phase inverter (without feedback), and
5. 5.3dB attenuation from negative feedback.

The power amp is at full power and on the cusp of overdrive. Upstream, the phase inverter is about halfway to overdrive in terms of peak voltage. For the first and second stages the situation depends on control positions. With the controls cranked up (orange trace), the second stage reaches overdrive before the first. With the controls at noon (blue trace), the first stage is closer to overdrive – the aggressive attenuation of the volume and tone control circuit creates a master-volume-like effect in which the upstream stage creates more distortion than the downstream stage.

With the controls at noon, input sensitivity at 1kHz is -12.6dBV (332mV peak), which is 13.1dB higher than for the 5E3 and 15.8dB higher than for the 6G3. With the volume cranked, it takes about 20dB less guitar to drive the amp to full power: -32.6dBV (33mV peak). With all knobs at maximum, sensitivity at 1kHz is -54.9dBV (3mV peak).

The first two stages operate well within their limits of headroom. The phase inverter has enough excess headroom to ensure that the power amp is well into overdrive before it begins to break up.

## System Voicing

With controls at 50-percent rotation (10-percent resistance) for the vibrato channel, here is the response from guitar input jack to the 6V6 grids normalized to the gain at 820Hz.

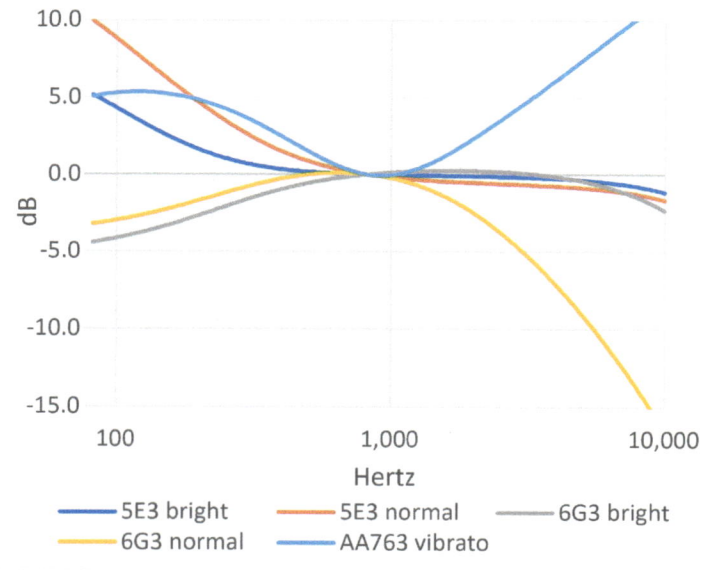

The AA763 introduces considerable middle scoop. Because the volume control is set to only 10-percent resistance, treble gets a big boost from the vibrato channel's bright bypass capacitor.

## Full Power Performance

The AA763 increases screen voltage and reduces DC grid bias.

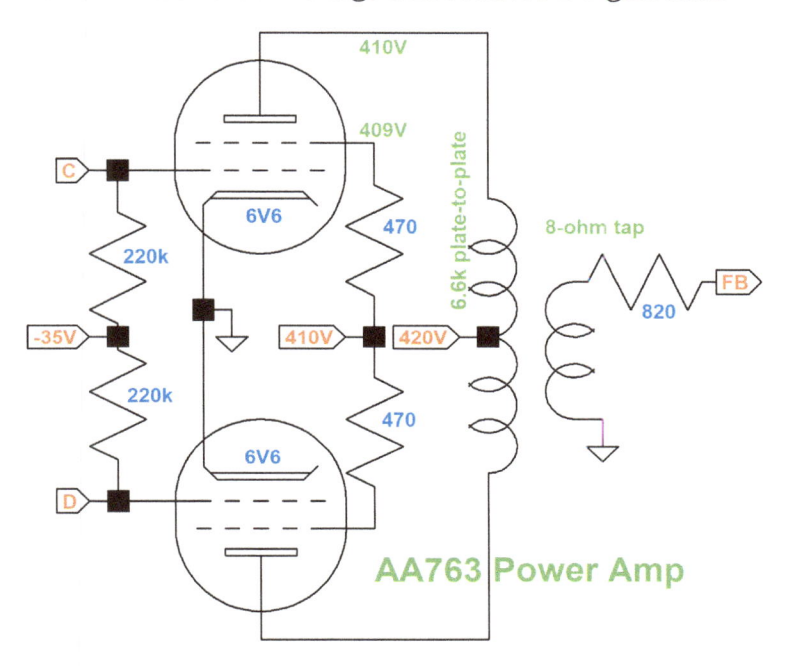

**AA763 Power Amp**

Here is a SPICE transient simulation[144] of one cycle of a 1kHz sine wave at full power.

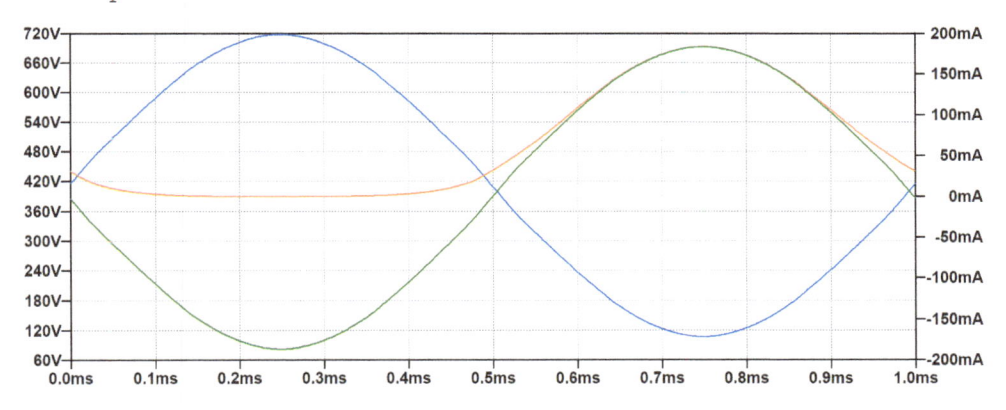

---

[144] Richard Kuehnel, **Guitar Amplifier Electronics: Circuit Simulation**, (Seattle: Amp Books, 2019).

The traces show upper 6V6 plate voltage (blue), upper 6V6 plate current (orange), and net plate current in the output transformer primary (green).

Plate current and screen current average 63mA and 10mA, respectively, for each tube.

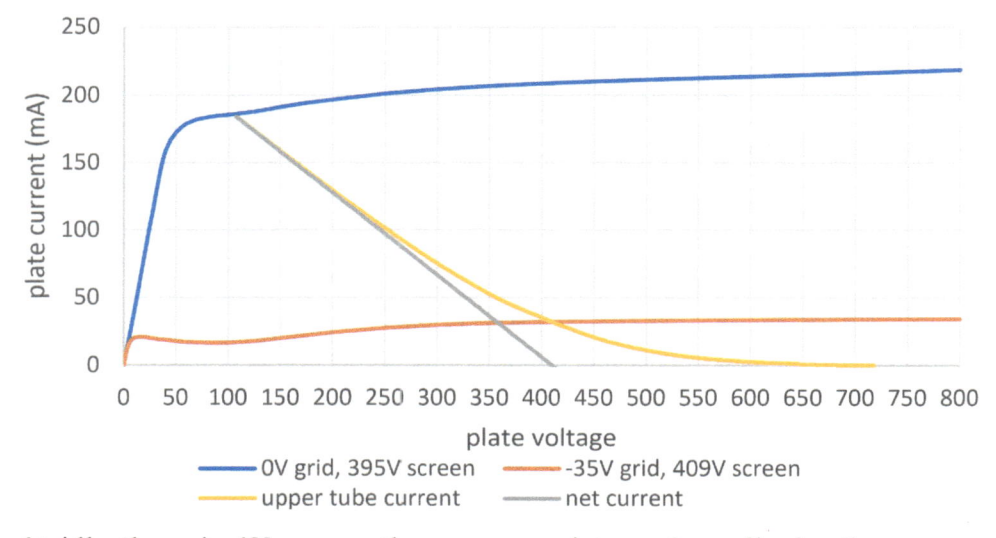

Here are the same performance curves we plotted for the 5E3 and 6G3.

*Chart: plate current (mA) vs plate voltage. Curves: 0V grid, 395V screen; -35V grid, 409V screen; upper tube current; net current.*

At idle there is 1V across the screen resistors. Accordingly, the orange curve is for a 409V screen and -35V grid. At peak screen current (32mA), the voltage across the 470Ω screen resistors reaches 15V, so the blue curve is for a 395V screen and 0V grid.

Upper tube plate current (yellow curve) swings down to only 0.2mA. With such a miniscule amount of electron flow, even Fender's marketing department must concede that the tubes reach cutoff – the AA763 is in Class AB operation at full power.[145] The plot shows that the upper tube reaches cutoff at high plate voltages and the net plate current equals the upper tube plate current at low plate voltages. The design's transition away from Class A and towards Class B becomes more pronounced as supply voltages sag.

The gray curve, representing the net plate current (upper tube plate current minus lower tube plate current) is 0mA when the plate voltage swings through its DC value of 410V. When the upper tube's grid swings to 0V, the plate voltage swings to 108V and the net plate current swings to 180mA. Full power is approximately

---

[145] Richard Kuehnel, **Guitar Amplifier Electronics: Basic Theory**, (Seattle: Amp Books, 2018), pp. 114-117.

$$\left(\frac{1}{2}\right)(410V - 108V)(180mA) = 27W$$

This is before supply voltage sag, which is greater for Class AB than for Class A. It also assumes an ideal output transformer and an 8Ω purely resistive speaker.

## Bias Excursion and Power Supply Sag

In overdrive, grid current causes the power amp's 0.1μF coupling capacitors to charge over the course of a few milliseconds. This makes the effective 6V6 DC grid bias more negative, creating bias excursion that pushes the tubes towards Class B operation.

Downstream of the GZ34 there is a total load current of 67.3mA under idle conditions. The power supply DC voltage is 420V, as marked by the red dot on the operation characteristics of a GZ34.

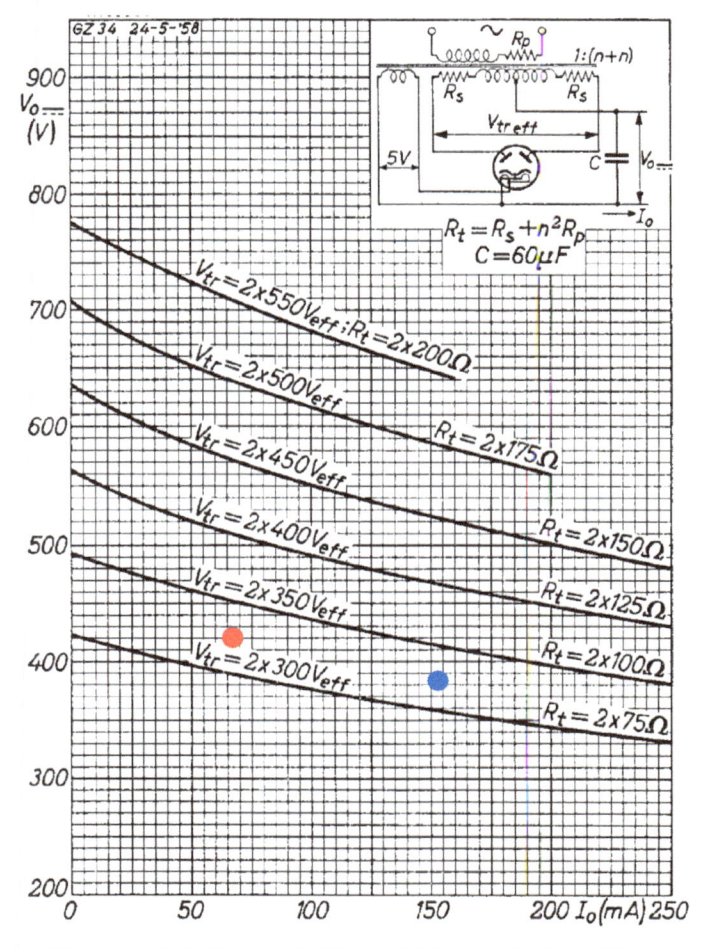

At full power, 6V6 plate current for two tubes averages 126mA. Screen current averages 20mA. Total power supply load current is 153.3mA, so the plate supply sags by about 37V, as shown by the blue dot. We saw a 23V sag for the 6G3, which operates closer to Class A with a warmer power amp bias. The AA763's colder bias demands less power supply current at idle but about the same at full power. The increased difference in load current between idle and full power translates to more sag.

These numbers assume, however, that average plate current and screen current hold steady over the course of the sag. In reality, full power load current decreases in response to lower supply voltages, so our dots represent boundaries for the power supply's operating range during a real musical performance. There is even more sag in overdrive.

## Modifications Incorporated into the AB763

The next iteration of the Deluxe incorporates several changes.

**tremolo**: The buffer is driven by a different node within the low-frequency oscillator circuit. Its cathode resistor value is almost doubled to 100kΩ.

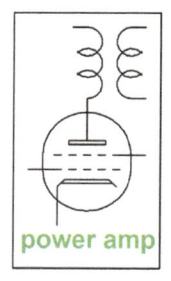

**power amp**: Grid-stopper resistors are added to filter radio frequencies and prevent radiofrequency oscillation. The resistors also improve the dynamics of overdrive and reduce the risk of blocking distortion.

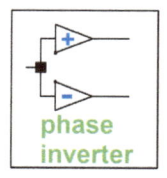

**phase inverter**: The tail resistor value is reduced to 22kΩ.

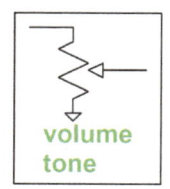

**volume tone**: The 0.033µF capacitor values in each stack are increased to 0.047µF.

## Tweaking the Tremolo, Power Amp, and Phase Inverter

The tremolo buffer is driven by the second high-pass RC filter in the low-frequency oscillator. The DC voltages shown here are Fender's measurements.

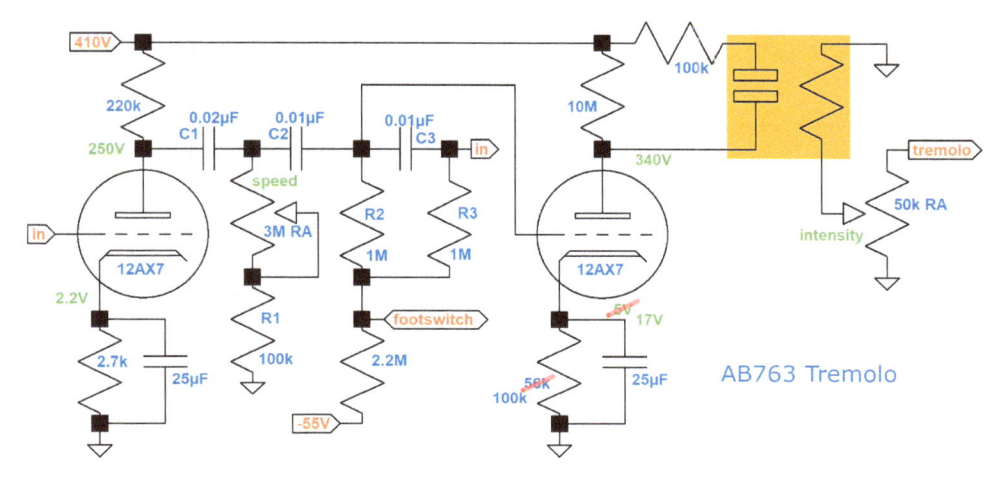

Each filter attenuates the signal, so the change in node increases the amplitude at the buffer input. The 56kΩ cathode resistor value for the buffer is increased to 100kΩ, which increases the DC cathode voltage to 17V.

For the power amp, the 1.5kΩ grid-stopper resistors and 6V6 inter-electrode capacitance create RC low-pass filters to block radio frequencies while allowing audio signals to pass.

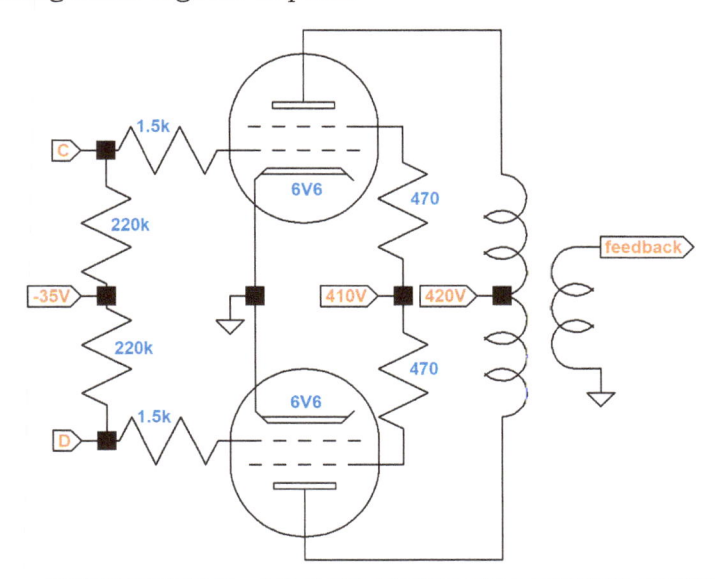

**PERFECTION – THE BLACKFACE DELUXE REVERB**

This keeps the amplifier from becoming a radio receiver or breaking into parasitic oscillation. Grid stoppers also improve the dynamics of overdrive by reducing the extent of bias excursion and shortening bias recovery time.[146]

The LTP 27kΩ tail resistor value is reduced to 22kΩ.

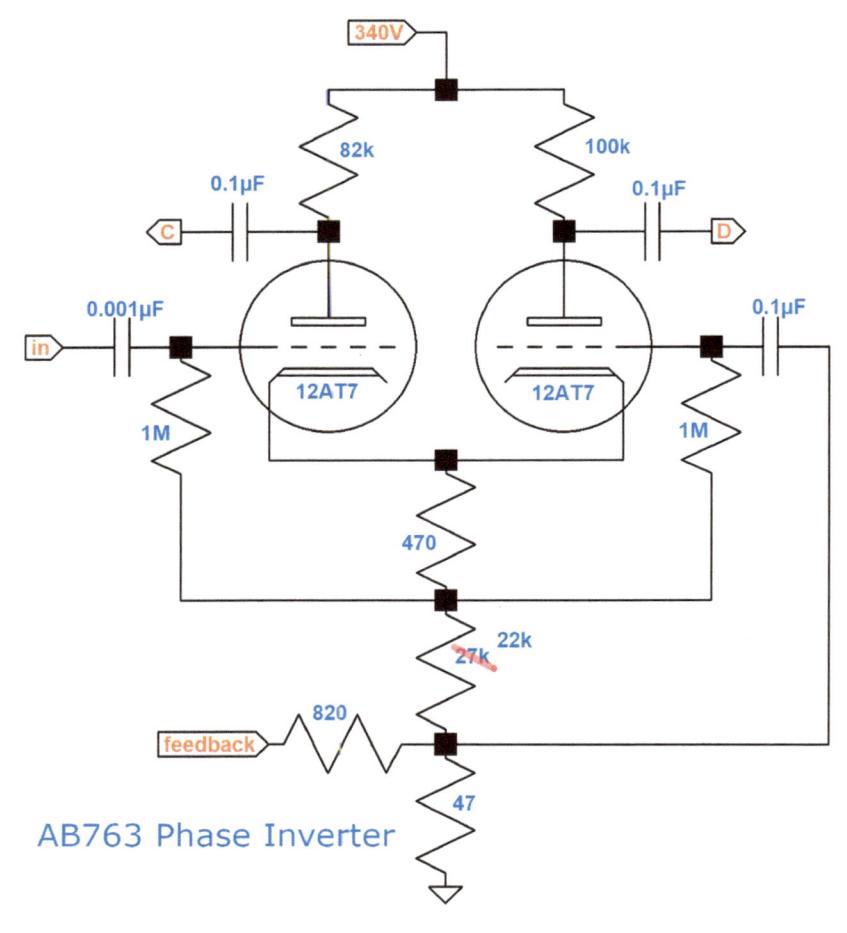

**AB763 Phase Inverter**

Decreasing the length of the tail increases maximum output voltage swing at the cost of increased imbalance between the two outputs, which increases 2nd-harmonic distortion. This is not a bad thing for a guitar amp.

On the next page is the response from the LTP input to the upper 6V6 grid (red, representing loaded gain for the inverted output) and the lower 6V6

---

[146] Richard Kuehnel, **Guitar Amplifier Electronics: Basic Theory**, (Seattle: Amp Books, 2018), pp. 166-170.

grid (blue, for the noninverted output). The plot on the left is for a 27kΩ tail and on the right is 22kΩ. The differences are inconsequential.

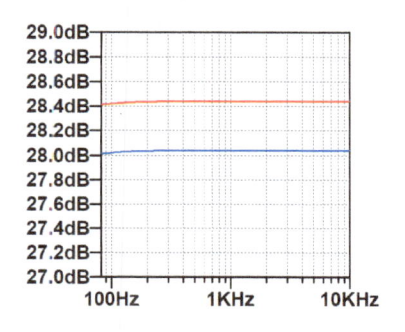

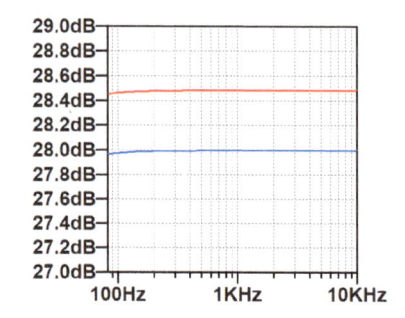

## Tweaking the Tone Stack

The AB763 increases the value of the tone stack's 0.033µF capacitor to 0.047µF. On the left is a SPICE AC analysis simulation of the AA763 response from stack input to the second stage grid with the volume control at maximum and the tone controls at minimum and maximum. On the right is the AB763 response.

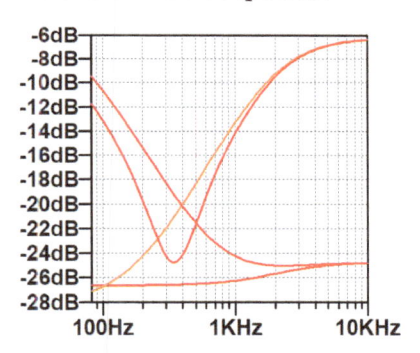

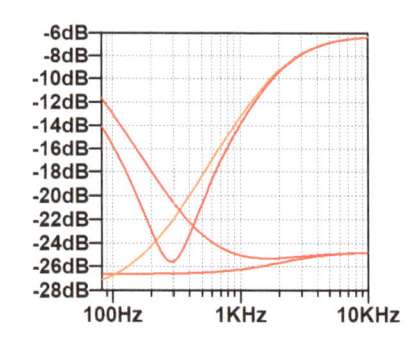

The capacitor modification shifts middle scoop to slightly lower frequencies. With all controls at 50-percent rotation (10-percent resistance), insertion loss at 1kHz is 44dB for the normal channel (below left) and 43.8dB for the vibrato channel (below right).

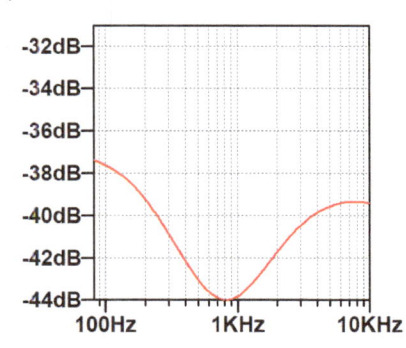

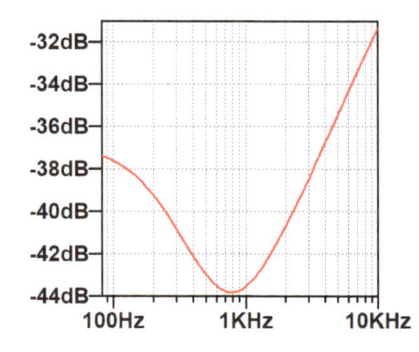

The plots assume the driving circuit output impedance is 38kΩ, which corresponds to the 100kΩ plate load resistor for the first-stage preamp in parallel with the 62.5kΩ plate resistance of a 7025/12AX7.

## Reverb Design Concept

To add reverb, Fender creates an alternate signal path. The normal channel is unchanged. The vibrato channel is divided into two signals: a *wet* signal that passes through the reverb circuit and a *dry* signal that bypasses reverb.

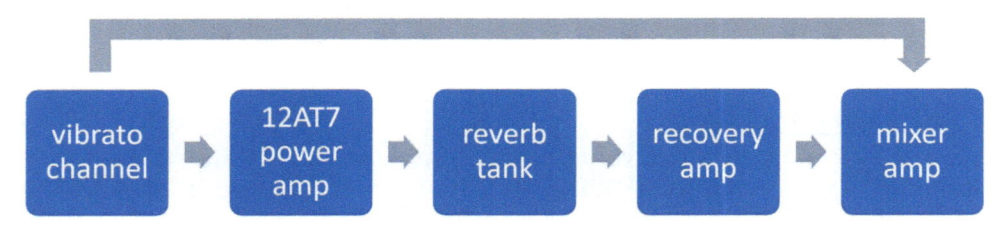

vibrato channel → 12AT7 power amp → reverb tank → recovery amp → mixer amp

The amount of wet signal in the mix is determined by a "reverb" control between the recovery amplifier and the mixer amplifier. The control represents an attenuating component, so its placement follows a familiar zigzag paradigm of guitar amplifier system design: a zig of amplification is followed by a zag of attenuation.[147] The wet signal undergoes a power amp zig, a reverb tank zag, a recovery amp zig, a reverb control zag, and a mixer amp zig.

The mixer amplifier and its coupling network affect only the vibrato channel.

> "This is subjective opinion, but adding that reverb circuit is a huge thing. Essentially you can think of it as a parallel effects loop. It changes the tone of the whole amp, rolling off the top in a pleasant way. There's still plenty of treble, but it's just different, with a musical warmth. You can hear all this readily."[148] –Steve Carr

The DC voltages shown in the schematics on the next three pages are Fender's measurements. There are 6 triodes connected to the lowest plate supply voltage, for which Fender does not provide a measurement. With some circuit analysis and a little high school algebra, however, we can estimate the voltage to be 275V. The calculation details are in the Appendix.

---

[147] Richard Kuehnel, **Fundamentals of Guitar Amplifier System Design**, (Seattle: Amp Books, 2019), pp. 16, 53, 139.
[148] Tom Wheeler, **The Soul of Tone**, (Milwaukee: Hal Leonard, 2007), p. 181.

# Deluxe AB763 Reverb

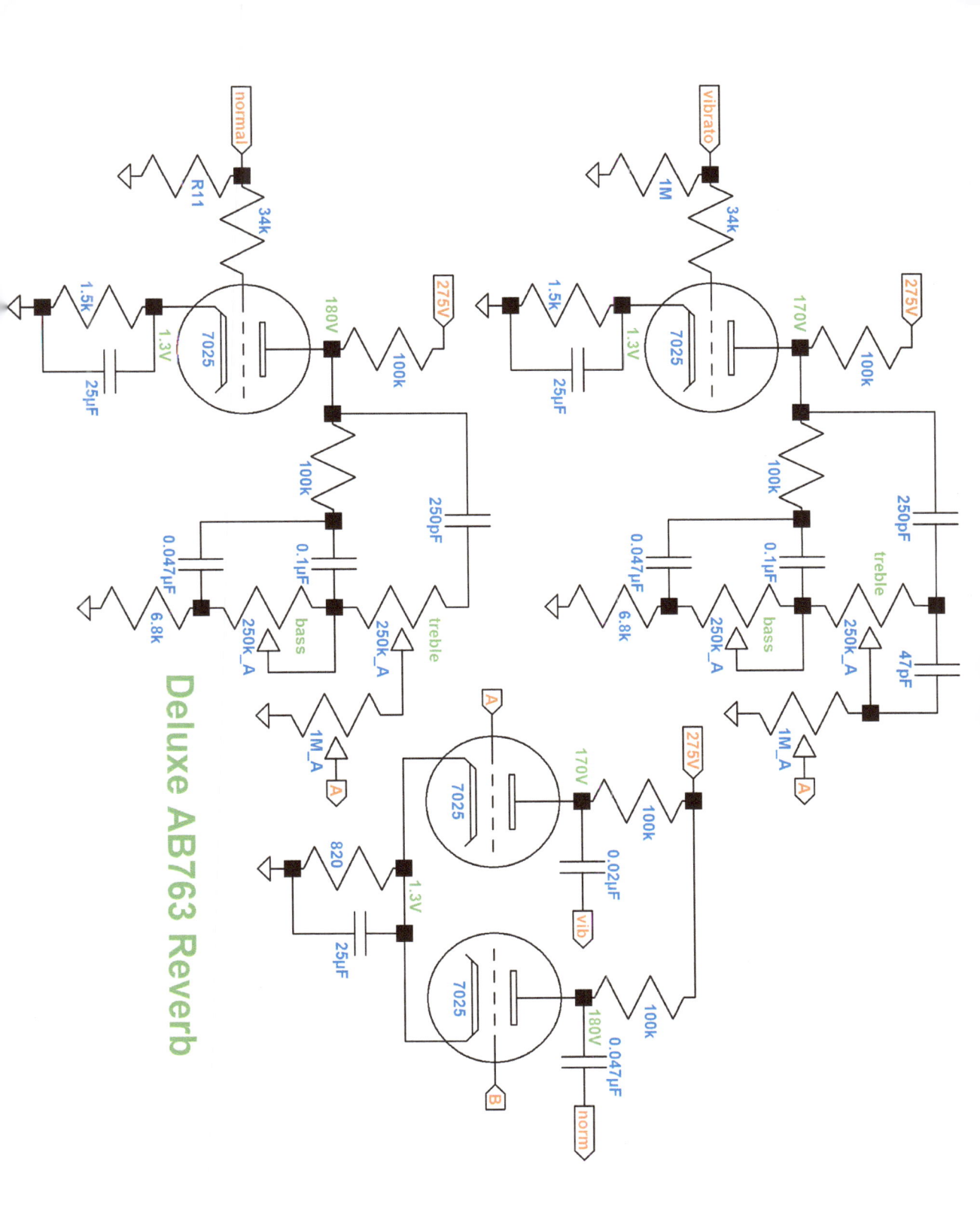

# Deluxe AB763 Reverb

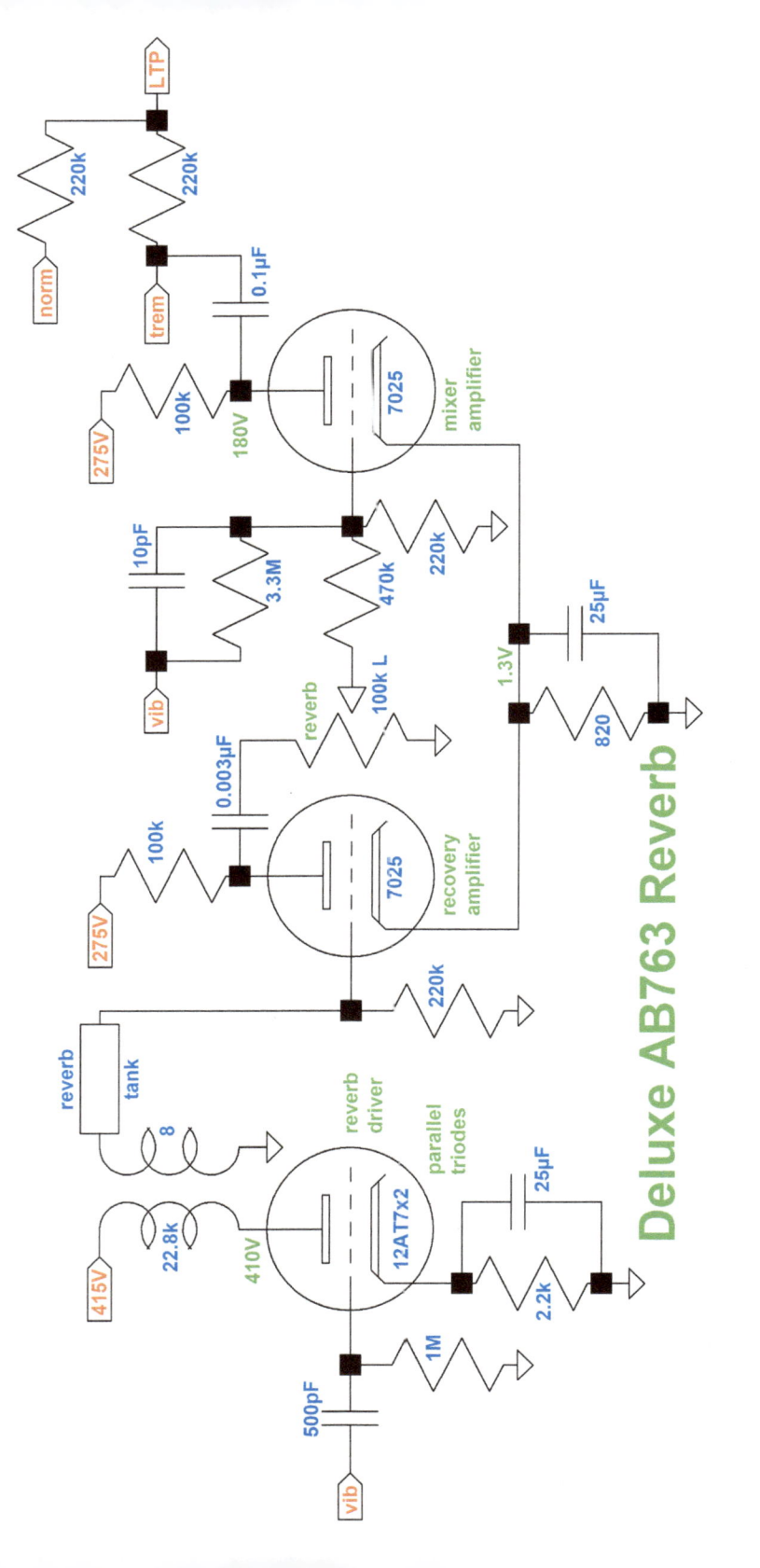

# Deluxe AB763 Reverb

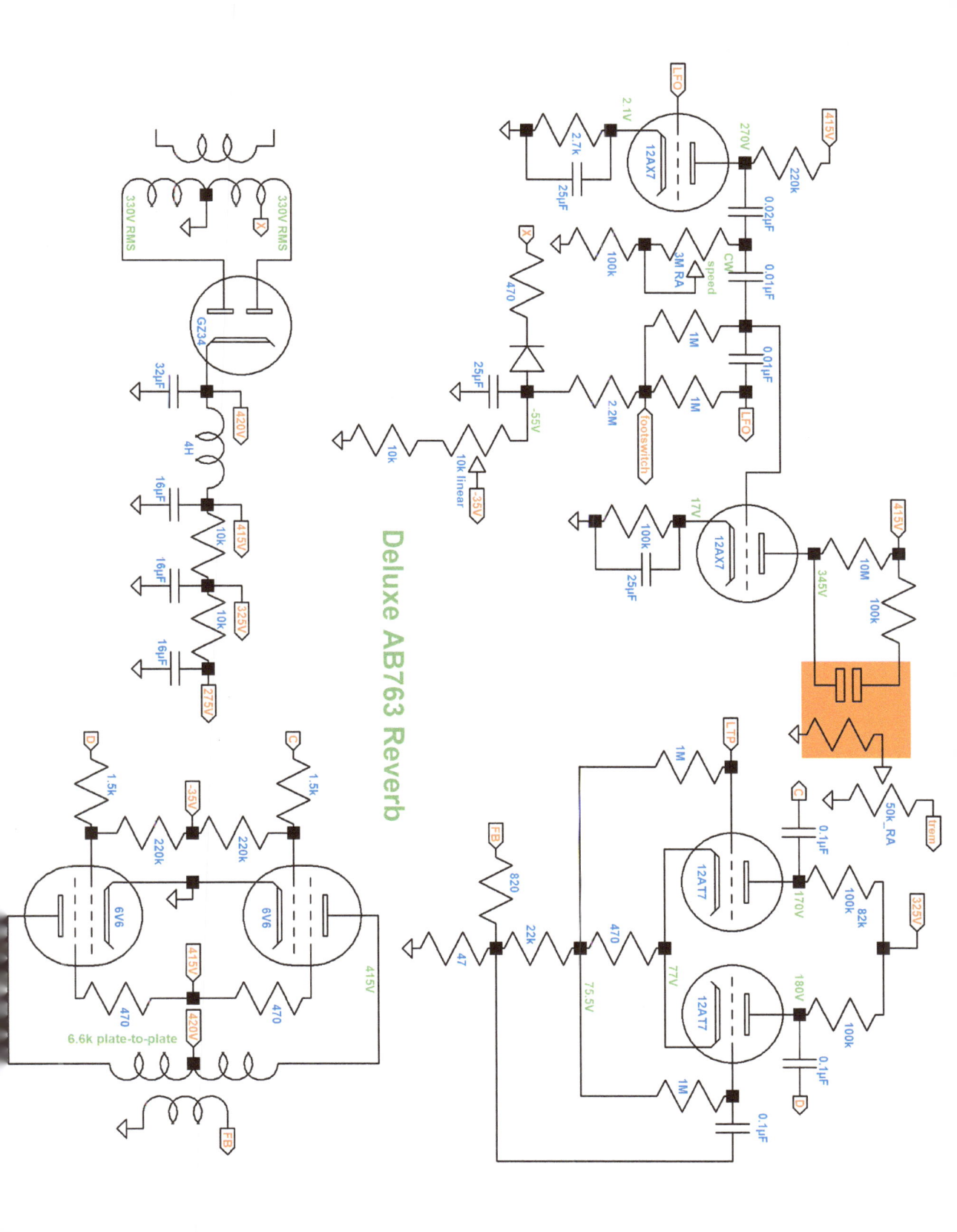

For the power supply, an additional 16μF capacitor is placed in parallel with the existing 16μF capacitor input for an effective capacitance of 32μF.

## Reverb Recovery and Mixer Amplifiers

Earlier we determined that with global negative feedback the phase inverter needs +5dBV (2.5V peak) at its input to drive the amp to full power. The output signal from the reverb tank is about 1 to 5 mV,[149] which is -63dBV to -49dBV. A 12AX7 voltage amplifier with a 100kΩ plate resistor and a fully bypassed cathode resistor has an unloaded gain of 35.8dB. Two voltage amplifiers in series give us 71.6dB, which is more than sufficient for the reverb tank signal to drive the amp to full power and beyond. The first is the *reverb recovery* stage and the second is the *mixer amplifier*, which amplifies the wet and dry signal mix.

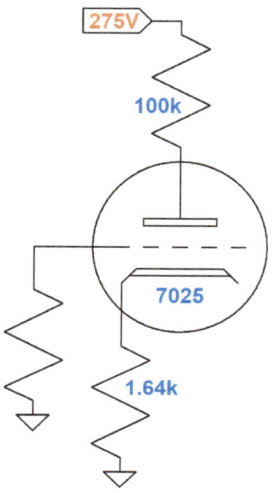

The recovery amplifier and mixer amplifier share a common 820Ω cathode resistor, so the equivalent for one triode is double: 1.64kΩ. The DC operating point is defined by -1.5V grid-to-cathode (Fender measures -1.3V), 178V plate-to-cathode (Fender measures 175V), and 0.9mA plate current, as shown by the red DC load line and the blue grid line on the next page.

The mixer amplifier drives a 220kΩ mixer network that is connected to the normal channel's second-stage output, which has a 38kΩ output impedance. If the normal channel is inactive, its output is at AC ground. Node `trem`, which represents the mixer amplifier output, drives a total AC load of 220kΩ + 220kΩ + 38kΩ = 478kΩ.

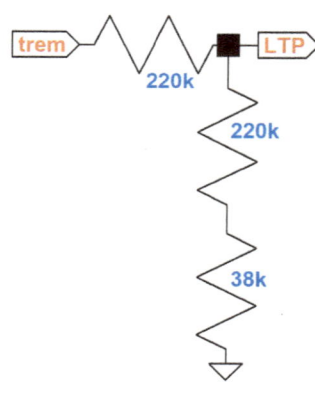

The AC plate load is thus equal to the mixer amplifier's 100kΩ plate load resistor and the 478kΩ AC load in parallel: 83kΩ. This corresponds to the slope of the green AC load line. The voltage "gain" of the 220kΩ mixer network is

$$\frac{220k\Omega + 38k\Omega}{220k\Omega + 220k\Omega + 38k\Omega} = 0.54 \, (-5.4dB)$$

---

[149] Amplified Parts, "Spring Reverb Tanks Explained and Compared," Available at https://www.amplifiedparts.com/tech-articles/spring-reverb-tanks-explained-and-compared (Retrieved January 14, 2021)

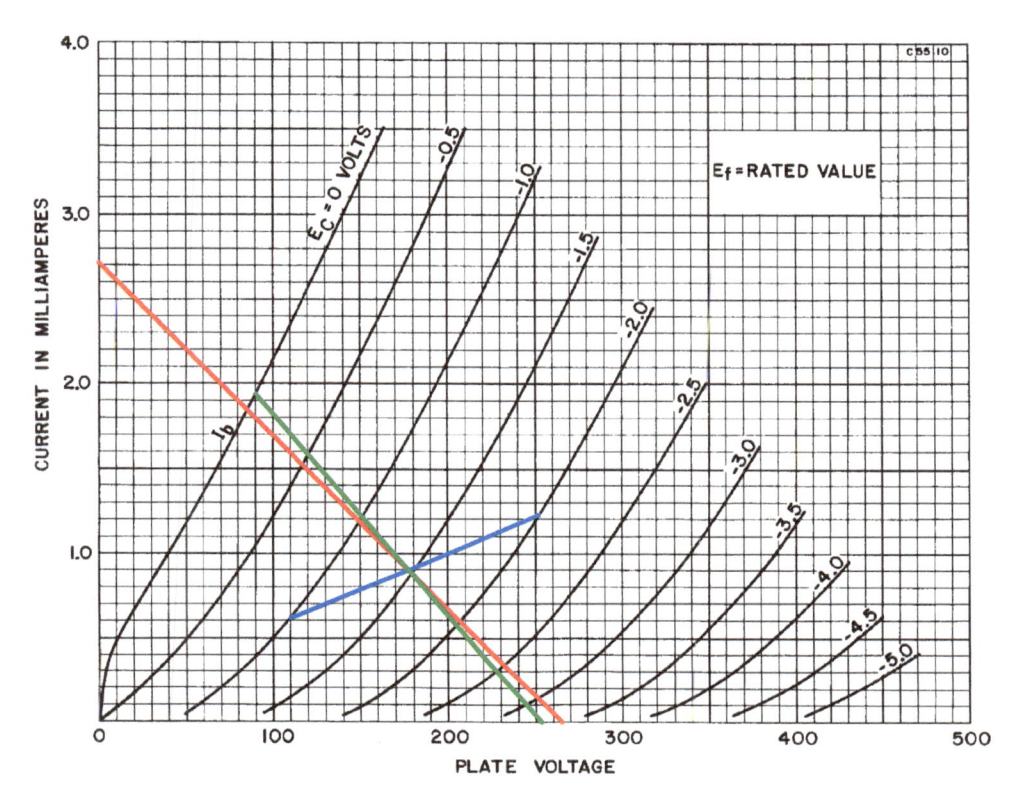

Loaded voltage gain for an amplification factor of 100, a plate resistance of 62.5kΩ, an 83kΩ AC plate load, and a fully bypassed cathode resistor is

$$\frac{(100)(83k\Omega)}{83k\Omega + 62.5k\Omega} = 57\ (35.1dB)$$

We determined earlier that the LTP phase inverter needs +5dBV at its input to drive the power amp to full power. The signal level needed at the mixer amp input is therefore

$$5dBV - 35.1dB + 5.4dB = -24.7dBV$$

Between the recovery amplifier and the mixer amplifier is the reverb control and an attenuator. The control has a linear taper, so at 50-percent rotation it is at 50-percent resistance. Downstream is a voltage divider formed by a 470kΩ series resistor and a 220kΩ shunt resistor, part values chosen to reduce the excess gain of two voltage amplifiers in series. The control represents a variable voltage divider that drives another voltage divider.

**PERFECTION – THE BLACKFACE DELUXE REVERB – 181**

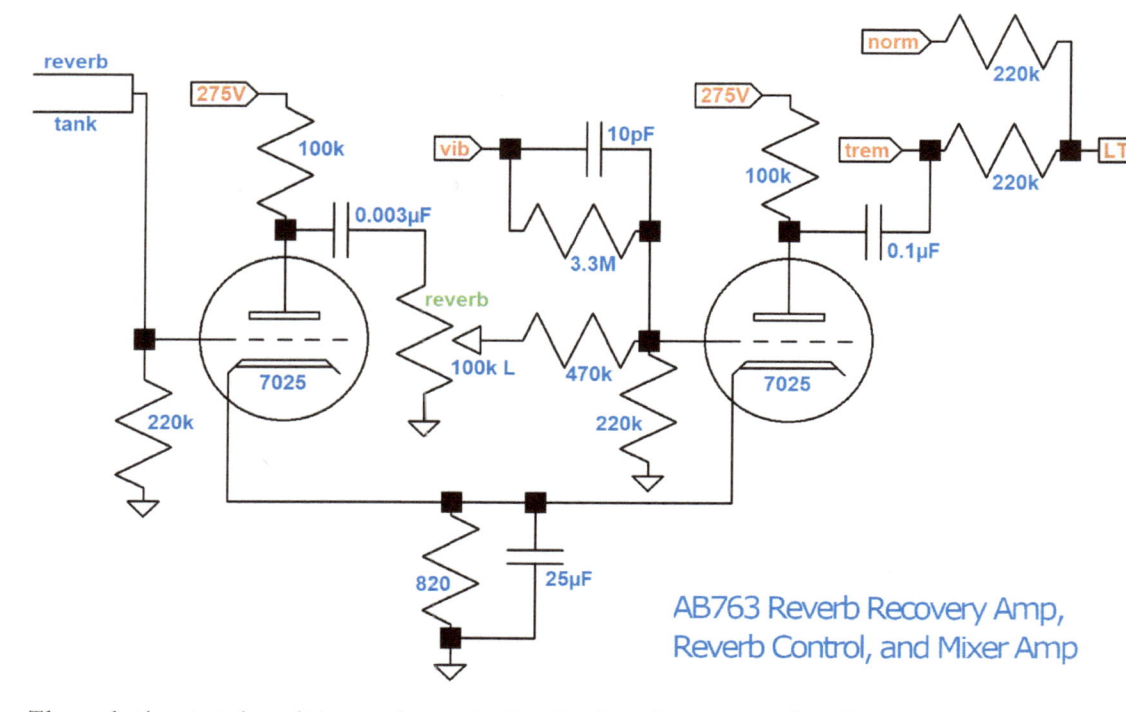

**AB763 Reverb Recovery Amp, Reverb Control, and Mixer Amp**

The relative total resistance in each divider has important implications:

1. The 470kΩ and 220kΩ resistors are much larger than the 100kΩ potentiometer, so the latter does not get significantly loaded by the former. This keeps the control fairly linear.
2. the 3.3MΩ resistor and 10pF capacitor have a resistance and reactance that is an order of magnitude greater than the 470kΩ and 220kΩ voltage divider, so they can be ignored for the purposes of determining the attenuation of the reverb signal.

When the reverb control is at 50-percent rotation (50-percent resistance for a linear taper), the equivalent circuit looks like this. The resistance between node A and ground is 50kΩ in parallel with $470k\Omega + 220k\Omega = 690k\Omega$. The latter is so much larger than 50kΩ that it can be ignored for determining the signal voltage at node A. The signal level at node A is therefore approximately half the voltage at the input, representing 6dB attenuation.

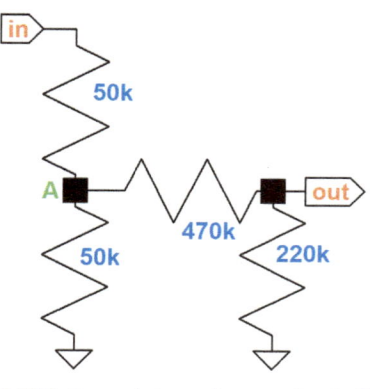

The voltage divider formed by the 470kΩ and 220kΩ resistors has a "gain" of

**182 – CHAPTER 5**

$$\frac{220k\Omega}{220k\Omega + 470k\Omega} = 0.32\ (-9.9dB)$$

The unloaded voltage gain for the recovery amplifier with an amplification factor of 100, a plate resistance of 62.5kΩ, a 100kΩ plate resistor, and a fully bypassed cathode resistor is

$$\frac{(100)(100k\Omega)}{100k\Omega + 62.5k\Omega} = 61.5\ (35.8dB)$$

The 0.003µF coupling capacitor is small considering the 100kΩ AC load. The -3dB break frequency for the load and a 38kΩ driving circuit output impedance is

$$\frac{1}{2\pi(100k\Omega + 38k\Omega)(0.003\mu F)} = 384Hz$$

At high frequencies for which the capacitor acts as a short circuit, the voltage divider formed by the 38kΩ output impedance and the approximately 100kΩ AC load creates a "gain" of -2.8dB at the top of the reverb control. The gain at 384Hz is less by 3dB: -5.8dB. Gain is -16.4dB at 82Hz, representing aggressive bass cut.

The relatively small capacitor value thus introduces significant bass attenuation, which is desirable when working with reverb. Reverberated bass can easily get muddy, as will be seen when we examine the characteristics of the reverb tank.

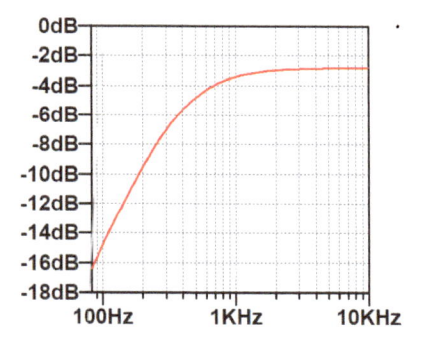

We determined that the mixer amplifier needs a signal level of -24.7dBV at its input for full power. The signal level needed at the reverb tank output to drive the amp to full power with the reverb control at noon is affected by the recovery amplifier unloaded gain, attenuation from its plate to the top of the reverb control, attenuation from the top of the control to node A, and attenuation by the voltage divider in front of the mixer amplifier:

$$-24.7dBV - 35.8dB + 2.8dB + 6dB + 9.9dB = -41.8dBV\ (11.5mV\ peak)$$

With the reverb control at maximum, the tank output needs to be about 6dB less: -47.8dBV (5.8mV peak). The system design is therefore well suited to the 1mV to 5mV expected from the tank. The wet signal is healthy but not likely to force the power amp into overdrive.

Let's take another look at the mixer amplifier from the perspective of the dry signal at node vib. Between node vib and the mixer amplifier grid is a voltage divider formed by a 3.3MΩ series resistor (with a 10pF bright bypass capacitor) and a shunt resistance of slightly less than 220kΩ. The "gain" is

$$\frac{220k\Omega}{220k\Omega + 3.3M\Omega} = 0.063\ (-24.1dB)$$

The dry signal gain from the vibrato channel's second-stage output to the LTP input is the "gain" of the attenuator plus the gain of the mixer amplifier plus the "gain" of the 220kΩ mixer network for the vibrato and normal channels:

$$-24.1dB + 35.1dB - 5.4dB = 5.6dB$$

By comparison, the gain for the normal channel's second-stage output to the LTP input is "gain" of the mixer network alone: -5.4dB. The vibrato and normal channels have different responses, even when the reverb control is set to minimum.

> "In my experience, most people use the Reverb channel even if they're not going to use the reverb. With its extra gain stage, it's almost like it has a bit more of a compelling sound."[150] –Steve Carr

## Reverb Tank

A reverb tank has a transducer at the input to convert the electrical signal into mechanical vibration, springs to convey the vibration from the tank input to the tank output, and a transducer at the output to convert the vibration back into an electrical signal. For the AB763 Fender licenses a Hammond Organ Company Type 4 reverb tank.[151]

> "The reverb circuit in Fender amps is very different from that in the company's three-knob outboard Reverb Unit of the early 1960s. The latter has more depth and 'sproing' and is undoubtedly the sound of surf - and maybe rockabilly-preferred tone as well - but many players prefer the smooth, round sound of the amp-based effect as typified on these blackface models."[152] –Dave Hunter

An Accutronics 4AB3C1B long-decay (41 milliseconds), 2-spring reverb tank is a suitable replacement for Fender's tank. According to its data sheet, the tank has these specifications:

- input impedance: 8Ω at 1kHz,
- output impedance: 2.25kΩ at 1kHz,

---

[150] Tom Wheeler, **The Soul of Tone**, (Milwaukee: Hal Leonard, 2007), p. 181.
[151] Tom Wheeler, p. 274.
[152] Dave Hunter, **Guitar Rigs: Classic Guitar & Amp Combinations**, (San Francisco: Backbeat Books, 2005), p. 157.

- typical decay time: 2.75 to 4.0 seconds,
- DC resistance: 0.81Ω, and
- Nominal drive current: 28.0mA RMS.

DC resistance is specified because it is easy to measure with a multimeter – a tank with a high input impedance also has a high input resistance. The impedance of a tank, however, is mostly inductive, so for circuit design we can assume that it is a pure inductive reactance[153] that is proportional to frequency. The 4AB3C1B has 8Ω of reactance at 1kHz, so at 2kHz, for example, the input impedance is 16Ω. At 500 Hz it is 4Ω.

The transducer input and output impedances at 1kHz are 8Ω and 2.25kΩ, respectively, so the transducer inductances are

$$\frac{8\Omega}{2\pi(1kHz)} = 1.27mH$$

$$\frac{2.25k\Omega}{2\pi(1kHz)} = 358mH$$

Unlike for a transformer, the low-inductance input and high-inductance output are not magnetically coupled. Instead, they are mechanically coupled via the tank's internal springs.

The output impedance is small in comparison to the 220kΩ grid-leak resistor that it drives, so the tank easily drives the recovery amplifier.

The 8Ω input impedance, on the other hand, is difficult to drive directly from a vacuum tube plate circuit. Fender's solution is the same as for a low-impedance speaker: the tank is driven by a power amplifier with an output transformer to transform the high plate circuit impedance to the low impedance of the AC load.

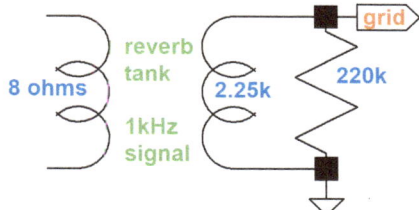

According to the tank's specifications, it needs a nominal drive current of 28mA RMS (39.6mA peak) to be fully driven. This means the voltage across the 8Ω tank input impedance needs to be

$$(39.6mA)(8\Omega) = 317mV\ peak$$

At 2kHz the impedance doubles and the required voltage swing doubles. At 10kHz it needs 3.17V peak to get the nominal drive current. With only 317mV peak at these higher frequencies there is less drive current, which is fine because high frequencies are not sonically pleasing for reverb. On

---

[153] Richard Kuehnel, **Guitar Amplifier Electronics: Basic Theory**, (Seattle: Amp Books, 2018), pp. 19-26.

the other hand, 317mV peak creates a drive current of 79.2mA at 500Hz and 396mA at 100Hz. In terms of frequency, to create a flat current response instead of a flat voltage response there needs to be aggressive low-frequency attenuation somewhere upstream of the tank. Otherwise, the reverb signal gets swamped by bass. Fender does indeed cut bass upstream, as we will see.

## Reverb Output Transformer

Our schematic shows a transformer primary impedance of 22.8kΩ, which corresponds to a Hammond 1750A replacement for Fender's 125A20B transformer.

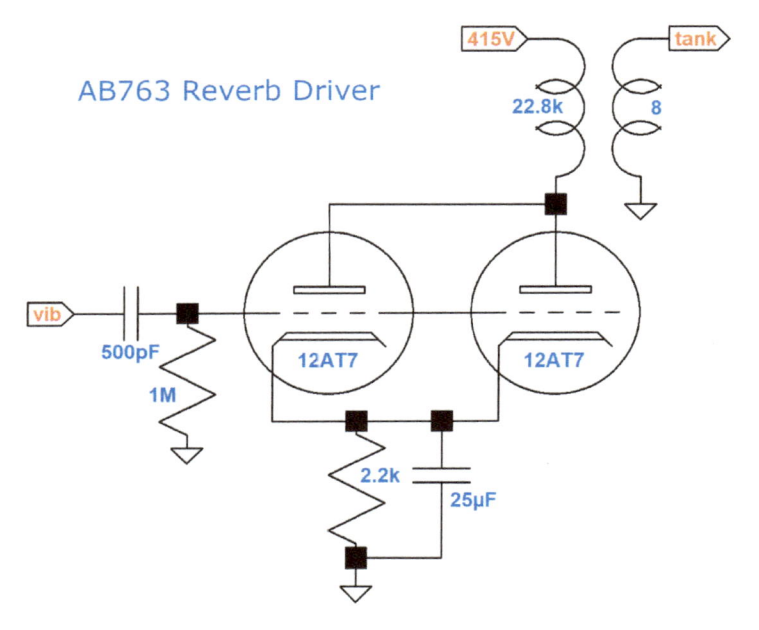

### AB763 Reverb Driver

(A Magnetic Components 40-18034 replacement transformer has a 25kΩ primary, a difference that is inconsequential.) For the 1750A, DC winding resistance is 1.065kΩ in the primary and 1.02Ω in the secondary.

| Turns ratio: | 53.11 : 1 |
|---|---|
| Inductance | @ 240V, 400Hz =>95.49H (Open circuit) |
| DCR | @ 20°C BLU - RED =1065Ω  ±20% |
|  | @ 20°C BLK - GRN =1.02Ω  ±20% |
| FREQ. RESP | @ 3.5W, 70Hz - 15kHz ±1dB  (1kHz Ref.) |
| Pri. Impedance | 22800Ω |
| Sec. Impedance | 8Ω |
| Power | 3.5 WATTS |

PRIMARY: BLU 22800Ω RED
SECONDARY: BLK 8Ω GRN

A 317mV peak voltage in the secondary translates to a peak voltage in the primary that is higher by a factor equal to the square root of the impedance ratio:

$$\sqrt{\frac{22.8k\Omega}{8\Omega}}(317mV) = 17V\ peak$$

(The square root of the impedance ratio approximately equals the turns ratio, for which Hammond specifies 53.11. Often, however, manufacturers do not include the turns ratio in their data sheets.)

## Parallel 12AT7 Reverb Driver

The reverb tank driver is a single-ended power amp with two 12AT7 triodes in parallel.

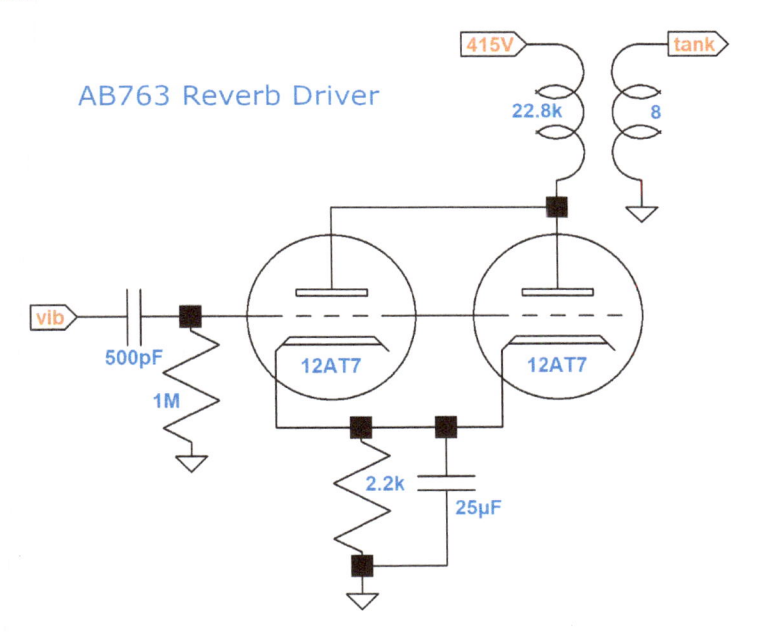

For DC the 25µF cathode capacitor is an open circuit and the inductance of the transformer is a short circuit. There is 2.2kΩ between the cathode and ground and, using Hammond's specification, 1.065kΩ DC winding resistance between the plate supply and the plate. They create a total resistance of 3.265kΩ in series with the tubes. The equivalent for one triode is double: 6.53kΩ.

If the plate current is 0mA, then the plate-to-cathode voltage is equal to the plate supply voltage: 415V. According to Ohm's Law, if the plate current increases by 30mA, then the plate-to-cathode voltage decreases by

$(30mA)(6.53k\Omega) = 196V$ to a value of 219V. These values mark the endpoints of the DC load line shown here in red.

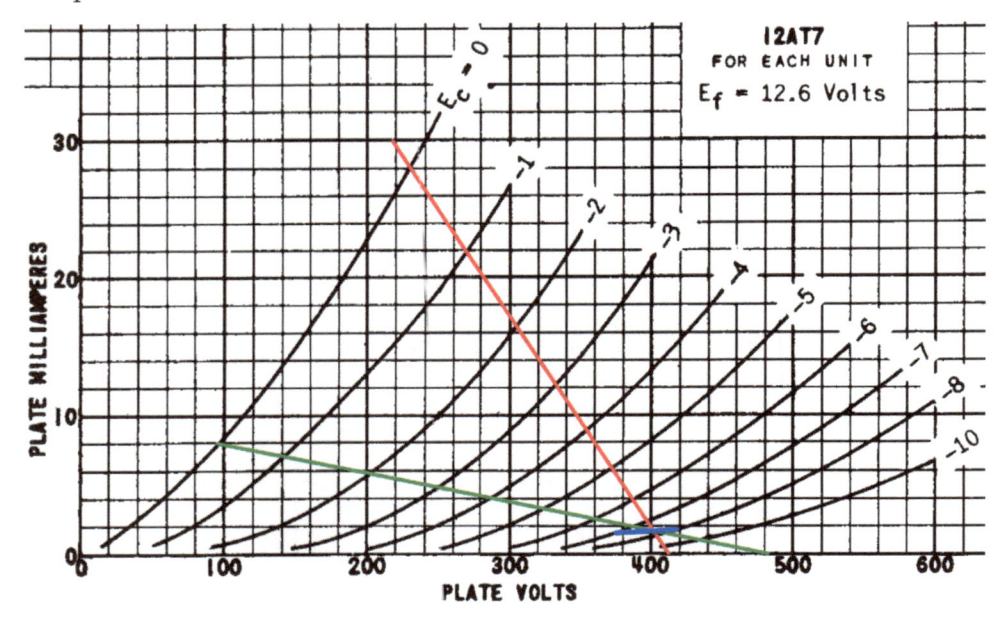

If the grid-to-cathode voltage is -7V or -8V, the plate current is

$$\frac{7V}{4.4k\Omega} = 1.6mA$$

$$\frac{8V}{4.4k\Omega} = 1.8mA$$

These are the endpoints of the blue line segment. The intersection is the DC operating point: a grid bias of -7.5V (Fender measures -8.7V), a plate current of 1.7mA, and a plate-to-cathode voltage of 401V (Fender measures 410V).

If the AC plate load for two triodes is 22.8kΩ, then the equivalent load for one triode is double: 45.6kΩ. When the plate current decreases from its DC value of 1.7mA to a value of 0mA, the plate voltage increases by $(1.7mA)(45.6k\Omega) = 78V$, so the plate voltage increases to 479V. This marks an endpoint of the green AC load line, which passes through the DC operating point.

According to the AC load line, a grid voltage swing of 2V peak is more than sufficient to create a plate voltage swing of 17V peak relative to the DC operating point. To drive the reverb tank to its nominal level at 1kHz, the signal level at the output of the second-stage voltage amplifier therefore can be about 2V peak (+3dBV).

**188 – CHAPTER 5**

The system profile for the AA763 (without reverb) indicates that the output of the second-stage voltage amplifier needs to be +5dBV (2.5V peak) to drive the power amp to full power.

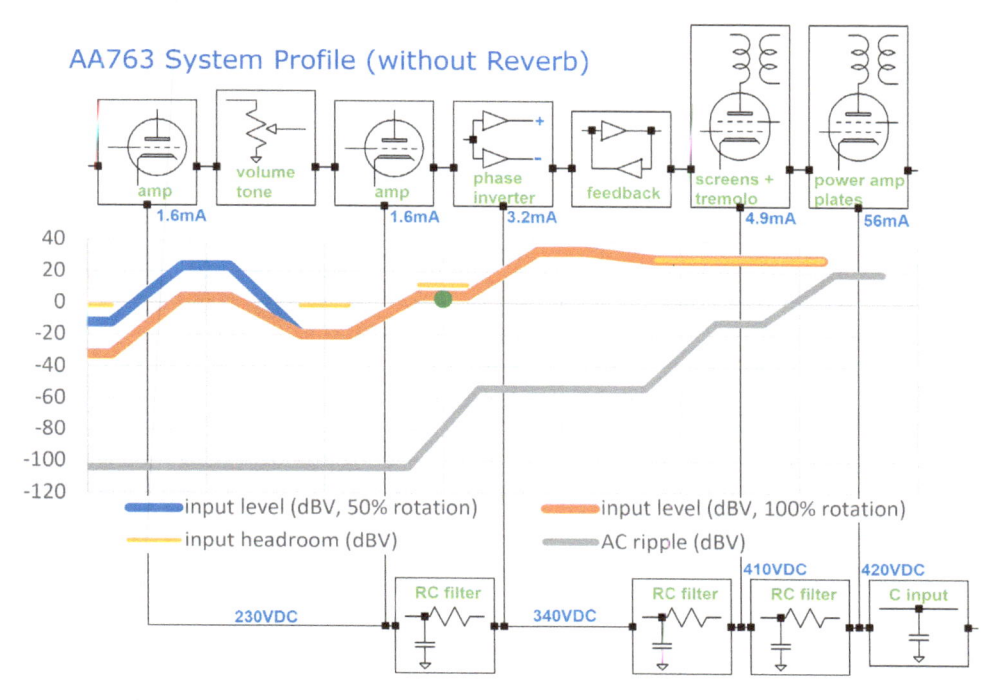

The added green dot is the signal level at that point for the AB763 Reverb, which is nearly the same. We can conclude that for a full-power 1kHz signal, the reverb tank is operating at close to its nominal signal level.

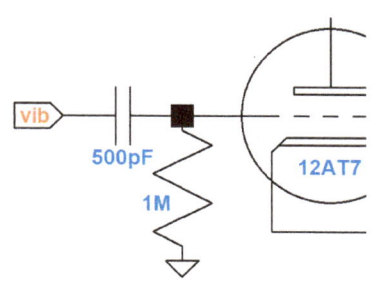

The 500pF capacitor at the input to the reverb driver is not for DC blocking – that is accomplished by an upstream 0.02µF coupling capacitor, as shown on the next page. As mentioned earlier, the input impedance to the reverb tank is almost purely reactive and proportional to signal frequency. Bass drives more current and can unpleasantly dominate the reverberated response. To mitigate this effect, Fender attenuates bass through the use of a relatively tiny, 500pF capacitor.

For the second-stage output impedance of 38kΩ and a 1MΩ grid-leak resistor, the -3dB break frequency is

$$\frac{1}{2\pi(1M\Omega + 38k\Omega)(500pF)} = 307 Hz$$

**PERFECTION – THE BLACKFACE DELUXE REVERB**

In other words, the resistance and the capacitor form a high-pass RC filter with a high break frequency. As the frequency decreases towards 82Hz, the lowest note on a guitar with standard tuning, gain decreases at a rate of 20dB per decade.

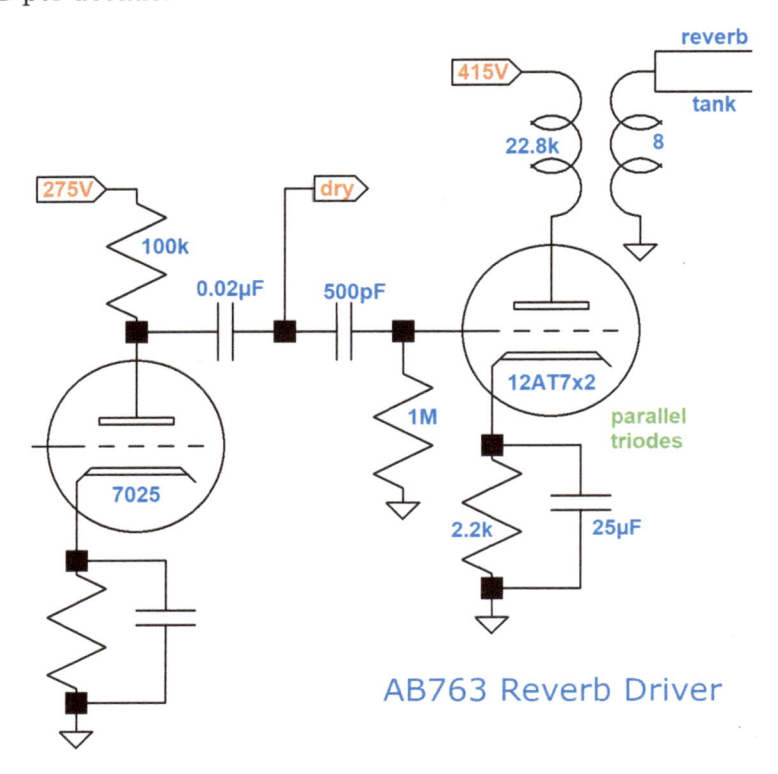

**AB763 Reverb Driver**

## Some Final Words on the AB763 Fender Deluxe Reverb

The amp has both throbbing tremolo and lush reverb. The 6V6 power tubes are driven to their limits, creating enough volume for stage and studio with just the right amount of breakup. The knob lineup is simple but effective, and with a grab-and-go weight under 50 pounds the amp is destined to become an indispensable tool for working musicians.

> "The Deluxe Reverb is a great amp for anything from blues to country to rock and roll, and even jazz. Consequently, it has long been a go-to combo for countless first-call Nashville and L.A. session players."[154] – Dave Hunter

> "Four 10s are great, but looking at all the factors, you've got the whole ball of wax in the Deluxe Reverb: a single 12, usually a pretty good one,

---

[154] Dave Hunter, "The Fender Deluxe Reverb," **Vintage Guitar**, June 2007.

well suited to the power of the amp, a very sophisticated front end with a good sound, that 6V6 cream – full of fat and cholesterol! – the reverb of choice, the right size, the right weight. If I had to have one amp to be stuck with on the desert island, give me a Deluxe Reverb. Turn it up, get a nice little breakup. Turn it down – sweet and clean!"[155] –Aspen Pittman

As Leo Fender contemplates the sale of his companies to CBS, the Deluxe reaches what many guitarists call *absolute perfection*.

---

[155] Tom Wheeler, **The Soul of Tone**, (Milwaukee: Hal Leonard, 2007), pp. 479-480.

# Appendix

## Decibel-Volts (dBV)

My *Basic Theory* book[156] covers fundamental electronics concepts and basic vacuum tube circuits. It does not, however, explain "decibel-volts," which are handy for evaluating the performance of multiple stages working together.[157] Just as decibels (dB) are convenient to describe the design of individual stages, decibel volts (dBV), representing a voltage level relative to 1V RMS, are convenient for system design. We use dBV throughout this book, both for computations and for plotting the results.

Peak voltage is greater than RMS voltage by a factor of the square root of 2, so to convert an audio signal level measured in volts peak to dBV, we divide by the square root of 2 and convert to decibels. A signal level of 100mV peak, for example, is -23dBV:

$$20 log \left(\frac{0.1V\ peak}{1.414V\ peak}\right) = -23 dBV$$

Let's say that a 100mV peak signal drives a preamp stage with a gain of 32dB followed by an attenuator with a "gain" of -6dB. The attenuator output signal level is a simple sum:

$$-23dBV + 32dB - 6dB = +3dBV$$

To convert the result back to volts peak, we divide the result by 20, take the inverse log, and multiply by the square root of 2:

$$(1.414V\ peak) log^{-1}\left(\frac{+3dBV}{20}\right) = 2V\ peak$$

Using simple sums expressed in dB and dBV is convenient for system design. It also makes it easier to plot signal levels.

## AB763 Supply Voltage Calculations

This is the procedure used to estimate the 275V plate supply voltage for the AB763 Reverb.

According to Fender's schematic, there are five triodes that draw current from the plate supply with an unknown voltage $V_{PP}$. Three have a 180V plate and two have a 170V plate. Each triode has a 100kΩ plate load resistor. According to Ohm's Law, the total current load is

---

[156] Richard Kuehnel, **Guitar Amplifier Electronics: Basic Theory**, (Seattle: Amp Books, 2018).
[157] Richard Kuehnel, **Fundamentals of Guitar Amplifier System Design**, (Seattle: Amp Books, 2019).

$$I_{PP} = 3\left(\frac{V_{PP} - 180V}{100k\Omega}\right) + 2\left(\frac{V_{PP} - 170V}{100k\Omega}\right) = \frac{5V_{PP} - 880V}{100k\Omega}$$

The plate supply voltage is equal to the 325V upstream supply voltage minus the voltage drop across the 10kΩ ripple filter resistor:

$$V_{PP} = 325V - (10k\Omega)I_{PP} = 325V - (10k\Omega)\left(\frac{5V_{PP} - 880V}{100k\Omega}\right) = 413V - (0.5)V_{PP}$$

We therefore conclude

$$V_{PP} = \left(\frac{413V}{1.5}\right) = 275.3V$$

As a check, three triodes have a DC plate current of

$$\frac{275.3V - 180V}{100k\Omega} = 0.953mA$$

Two have a DC plate current of

$$\frac{275.3V - 170V}{100k\Omega} = 1.053mA$$

for a total current of 4.97mA. The voltage drop across the 10kΩ ripple filter resistor is therefore 49.7V, creating a plate supply voltage of $325V - 49.7V = 275.3V$.

# Index

AC load line, 51, 87
AC ripple, calculating, 43
Bair, Mark, 130
bias
  cathode, 3, 8
  grid-leak, 27
Black, Bill, 34
Bonamassa, Joe, 141
break frequency
  coupling capacitor, 13
Buchanan, Roy, 2
Campbell, Mike, 91, 98
Carr, Steve, 8, 176
cathode degeneration, 114
cathode follower, 94
cathode-heater voltage, 68
choke, 149
clamping, 93
Class A operation, 82, 96, 170
Class AB operation, 85, 137, 148, 170
Class B operation, 83, 95, 96, 133, 138, 170
coupling capacitor
  break frequency, 13
Cragg, Larry, 2
Cunningham, Madison, 2
current
  grid, 28
damping factor, 87, 88
Danelectro, 101
dBV, 192
DC grid bias, 28
DC load line, 46
decibel volts (dBV), 192
decoupling
  plate circuit, 10
distortion
  blocking, 96, 157
  harmonic, 84, 88
  intermodulation (IMD), 88
  nonlinear, 89
  split-load phase inverter, 91
  total harmonic (THD), 84
Dr. Z Amplification, 2
Dr. Z DB4, 2
evolutionary development, 2, 55
feedback and control systems, 115

Fender
- Bandmaster, 69, 77, 101, 141
- Bassman, 40, 69, 109, 141
- Concert, 124, 141, 144
- Princeton, 5, 20, 39, 77
- Pro, 20, 38, 39, 69, 101, 141, 144
- Professional, 5
- Showman, 141, 144
- Super, 3, 20, 38, 39, 69, 101, 141, 144
- Tremolux, 101, 141, 144
- Twin, 35, 38, 39, 57, 59, 69, 101, 109, 141
- Vibrasonic, 84, 124, 141
- Vibrolux, 141, 144
- Vibroverb, 141, 144

Fender, Leo, 5, 8, 60, 71, 84, 99, 114, 159, 191

filter
- graded, 9
- ripple, 10

gamma network, 77
Gibbons, Billy, 91
Gibson, 101
graded filter, 9
grid bias, 28
grid current, 28
grid-leak bias, 27
grid-stopper resistor layout, 62
Heartbreakers, 1, 98
Hunter, Dave, 11, 22, 57, 80, 83, 88, 92, 119, 184, 190
Huntington, Greg, 145

impedance
- output, 12
- speaker, 85

input sensitivity, 19
insertion loss, 16
Jensen 12R-8Ω speaker, 87
Letts, Mike, 99

load line
- AC, 51, 87
- DC, 46

logarithmic taper, 14

long-tailed-pair phase inverter
- how it works, 109
- imbalance, 110

low-frequency oscillator (LFO), 130
maintaining voltage (for a neon lamp), 153
Mason, Victor, 98
Massie, Ray, 8
Miller capacitance, 76
Moore, Scotty, 1, 31, 34

motorboating, 10, 106
Multivox Premier, 101
National Association of Music Merchants (NAMM), 35, 162
negative feedback, 40, 91
    Deluxe 5C3, 38
    Deluxe 6G3, 114
    Deluxe AA763, 157
    effect on overdrive, 91
    formula for, 115
neon lamp, 150
    normal glow region, 153
Nicholas, Shane, 78, 88, 107, 114
output impedance, 12
Paisley, Brad, 2
paraphase phase inverter
    floating, 44
    how it works, 11
    self-balancing, 44
Peavey, Hartley, 7, 99
Petty, Tom, 1
Pittman, Aspen, 2, 190
plate circuit decoupling, 10
potentiometer
    CTS Electrocomponents Series 450G, 14
    taper, 14
power chord, 90
presence control, 115
Presley, Elvis, 1, 34
Radio & Television Equipment Company, 8
Randall, Don, 8, 40
RC ripple filter, 44
Reid, Paul, 4, 11, 96
resistor
    grid-stopper, 8
    light-dependent (LDR), 150
    screen, 8, 145
resonance
    speaker, 88
reverse-audio taper, 152
Rickenbacker, 39
    grid-leak bias, 28
ripple filter, 10
    rule of thumb for graded filters, 106
Robinson, Rich, 91
sensitivity, 19
signal, wet and dry, 176
Smith, Richard, 5
speaker
    impedance, 85
    resonant frequency, 87

split-load phase inverter, 65
system profile, 53
system voicing, 55
Teagle, John, 130
Tedeschi, Susan, 2
Thomas, Bob, 57, 71, 90
transformer impedance, 83
tremolo
  bias-modulation, 129
  grid-bias, 129
  low-frequency oscillator (LFO), 130
  signal-modulation, 150
tremolo versus vibrato, 129
trickle-down economics, 141
vacuum tube
  12AX7, 25, 44
  12AX7, parameters for low DC plate current, 128
  12AY7, 48
  12AY7, plate characteristics, 49
  5879, 3
  5Y3, 7
  5Y3, plate characteristics, 106
  6J5, 35
  6N7, 10, 11, 25
  6SC7, 7, 11, 24
  6SF5, 5
  6SL7, 10
  6SN7, 11
  6V6, plate characteristics, 42
  GZ34, operation characteristics, 139
vibrato versus tremolo, 129
voltage gain
  attenuator, 12
  closed-loop, 115, 118, 159
  feedback, 115, 118, 159
  forward, 115
  loaded, 113, 161
  open-loop, 115
  unloaded, 12
  voltage divider, 12
Western Electric, 7
Wheeler, Tom, 5, 45, 60, 71, 99, 124, 141, 159
Wilkens, Matt, 31, 130, 163
Young, Neil, 2, 91
Zaite, Mike, 3